PRINCIPLES OF
WOVEN FABRIC
MANUFACTURING

PRINCIPLES OF WOVEN FABRIC MANUFACTURING

ABHIJIT MAJUMDAR

CRC Press
Taylor & Francis Group
Boca Raton London New York

CRC Press is an imprint of the
Taylor & Francis Group, an **informa** business

CRC Press
Taylor & Francis Group
6000 Broken Sound Parkway NW, Suite 300
Boca Raton, FL 33487-2742

First issued in paperback 2020

ISBN 13: 978-0-367-57419-2 (pbk)
ISBN 13: 978-1-4987-5911-3 (hbk)

ISBN-13: 978-1-4987-5911-3 (hbk)

Library of Congress Cataloging-in-Publication Data

Names: Majumdar, Abhijit, author.
Title: Principles of woven fabric manufacturing / Abhijit Majumdar.
Description: Boca Raton : CRC Press, 2017. | Includes bibliographical references and index.
Identifiers: LCCN 2016025821 | ISBN 9781498759113 (hardback : acid-free paper)
Subjects: LCSH: Weaving--Textbooks.
Classification: LCC TS1490 .M327 2017 | DDC 677/.028242--dc23
LC record available at https://lccn.loc.gov/2016025821

Visit the Taylor & Francis Web site at
http://www.taylorandfrancis.com

and the CRC Press Web site at
http://www.crcpress.com

Contents

Preface

THE ART OF MAKING fabric through weaving is as old as human civilisation. Why should students of the twenty-first century show interest in such an old technology? Well, in ancient times, pigeons were used as messengers. Then came postmen, who delivered letters from different countries to our abode. And today, 'communication engineering' has become one of the most vibrant and sought-after engineering branches. Therefore, as far as interest creation is concerned 'how you do' is probably more important than 'what you do'. In the same note, the objective of weaving has still remained the same: to produce fabrics. However, the way of producing fabrics has transformed in the last three centuries. With the advancement in machine design, electronics and automation, weaving has become a very sophisticated technology.

The objective of this book is to introduce the basic concepts of woven fabric manufacturing to the undergraduate students of textile engineering. Other popular fabric manufacturing technologies like knitting, nonwoven, and braiding are not included in this book. The book covers the weaving technology based only on scientific and engineering approaches. The author has attempted to introduce the fundamental aspects of weaving preparatory and weaving assuming that the students have knowledge of high-school mathematics and science. Some mundane process descriptions which require memorising have been deliberately avoided. This pithy book does not cover industrial practices of weaving. It essentially covers the 'vital few' and not the 'trivial many'.

The content has been designed to create interest among students to hone their analytical abilities. The author has conviction that after reading this book, students will be able to understand and analyse the preparatory processes of weaving such as winding, warping and sizing. They will also be able to analyse various mechanisms of shuttle and shuttleless looms such as shedding, picking, beat-up, take-up and let-off.

Most of the mathematical equations presented in this book have been derived from first principles so that the readers can comprehend them better and understand their implications. Topics such as shedding cam design and sley kinematics have been discussed in detail as students find it interesting to study these engineering aspects of weaving technology. Each chapter contains some solved numerical problems to enhance clarity of the readers about the topic discussed. Many complex situations of weaving preparatory and weaving have been explained with simple hypothetical analogies. These analogies may not always represent the prevailing complexities; however, they are essential for concept building.

There are a few topics which should have been discussed in greater depth. The author feels that water-jet weaving is one of those and wishes to improve on this topic in subsequent editions. In spite of best efforts, there might be some inadvertent mistakes in the book. It will be appreciated if the readers can spot these errors and bring it to the author's notice.

Abhijit Majumdar

Acknowledgements

THE AUTHOR EXPRESSES HIS gratitude to Prof. Prabir Kumar Banerjee and Prof. Vijay Kumar Kothari of Indian Institute of Technology Delhi for introducing him to the world of fabric manufacturing. He is highly obliged to Dr. Abhijit Biswas of the Government College of Engineering and Textile Technology, Berhampore, India, and Prof. M. Madhusoothanan of Anna University, Chennai, India, for providing valuable comments and suggestions to improve the contents of the book. The author is also thankful to Prof. Rabi Chattopadhyay of IIT Delhi for providing inspiration to pursue this initiative. A special mention must be made of Dr. Anindya Ghosh and Dr. Tarit Guha, of Government College of Engineering and Textile Technology, Berhampore, India, with whom the author had endless discussions and exchanges while preparing this manuscript. Thanks also to Khashti Pujari and Animesh Laha for their help in preparing some of the diagrams of this book. The author gratefully acknowledges permissions accorded by M/S Prashant Industries, India, M/S Uster Technologies, Uster, Switzerland, CRC Press in Boca Raton, Florida, Elsevier, Amsterdam, Netherlands, and The Textile Institute in Manchester, United Kingdom for the use of copyrighted figures in this book. Finally, the author thanks Dr. Gagandeep Singh of CRC Press wholeheartedly for his patience and perseverance in following up preparation of the manuscript.

Author

Abhijit Majumdar obtained bachelor's degree from Calcutta University in 1995 with first class first position in Textile Technology program. He acquired master's degree in textile engineering from Indian Institute of Technology (IIT) Delhi in 1997 and doctorate in engineering from Jadavpur University, Kolkata, in 2006. He also holds an MBA from IIT Delhi with specialization in operations management.

He worked in companies such as Voltas Limited and Vardhman Group before joining academia. He taught in Government College of Engineering and Textile Technology, Berhampore, West Bengal, India, between 1999 and 2007. He joined IIT Delhi as assistant professor in 2007. At present, he is working as associate professor in Textile Engineering Group. His research areas include protective textiles, soft computing applications and operations and supply chain management. He has completed three research projects funded by the Department of Science and Technology (DST) and Council for Scientific and Industrial Research (CSIR), India. He has published 70 research papers in international refereed journals and guided four PhD students. He has edited two books published by Woodhead Publisher, U.K., and authored one monograph (*Textile Progress*) published by Taylor & Francis Group.

He is the associate editor of the *Journal of the Institution of Engineers (India) Series E (Chemical and Textile Engineering)* published by Springer. He is a recipient of Outstanding Young Faculty Fellowship of IIT Delhi (2009–2014) and the Teaching Excellence Award (2015) of IIT Delhi.

CHAPTER 1

Introduction to Fabric Manufacturing

1.1 FABRIC MANUFACTURING TECHNOLOGIES

Textile fabrics are generally two-dimensional flexible materials made by inter-lacing of yarns or intermeshing of loops with the exception of nonwovens and braids. Fabric manufacturing is one of the four major stages (fibre production, yarn manufacturing, fabric manufacturing and textile chemical processing) of textile value chain. Most of the apparel fabrics are manufactured by weaving, though knitting is catching up or even moving ahead very fast, especially in the sportswear segment. Natural fibres in general and cotton fibre in particular are the most popular raw material for woven fabrics intended for apparel use. Staple fibres are converted into spun yarns by the use of a series of machines in the yarn manufacturing stage. Continuous filament yarns are often texturised to impart spun yarn-like bulk and appearance.

Textile fabrics are special materials as they are generally light-weight, flexible (easy to bend, shear and twist), mouldable, permeable and strong. The four major technologies of fabric manufacturing are as follows:

1. Weaving

2. Knitting

3. Nonwoven

4. Braiding

Figure 1.1 depicts the fabrics produced by the four major technologies.

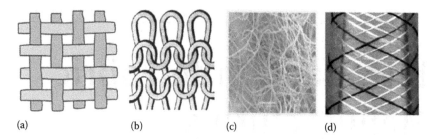

(a) (b) (c) (d)

FIGURE 1.1   Fabrics produced by different technologies. (a) Weaving, (b) knitting, (c) nonwoven and (d) braiding.

TABLE 1.1   Properties of Some Technical Fabrics

Fabric Type	Important Properties/Parameters
Filter fabrics	Pore size, pore size distribution
Body armour fabrics	Impact resistance, areal density, bending resistance
Fabrics as performs for composite	Tensile strength and tensile modulus
Knitted compression bandages	Stretchability, tensile modulus, creep

Fabric manufacturing may be preceded either by fibre production (in case of nonwoven) or by yarn manufacturing (in case of weaving, knitting and braiding). Fabrics intended for apparel use must fulfil multidimensional quality requirements in terms of drape, handle, crease recovery, tear strength, air permeability, thermal resistance and moisture vapour permeability. However, looking at the unique properties and versatility of textile fabrics, they are now being used in various technical applications where the requirements are altogether different. Some examples are given in Table 1.1.

1.2 WEAVING TECHNOLOGY

Weaving is one of the oldest technologies of human civilisation and has been in existence since 7000 BC. The weaving technology leapfrogged with the invention of power loom by Edmund Cartwright in 1785 (Lord and Mohamed, 1982) and is considered to be one of the key inventions of the Industrial Revolution (1760–1840). Weaving is the most popular method of fabric manufacturing and is generally done by interlacing two orthogonal sets of yarns – warp (singular: end) and weft (singular: pick) – in a regular and recurring pattern. Actual weaving process is preceded by yarn preparation processes, namely winding, warping, sizing, drawing and denting.

Winding converts the smaller ringframe packages to bigger cheeses or cones while removing the objectionable yarn faults. Pirn winding is

FIGURE 1.2 Types of yarn packages. (a) Ringframe bobbin or cop, (b) cone, (c) cheese and (d) pirn.

performed to supply the weft yarns in shuttle looms. Figure 1.2 shows various yarn packages used in textile operations (from left to right: ringframe bobbin or cop, cone, cheese and pirn). Warping is done to prepare a warper's beam which contains a large number of parallel warp yarns or ends in a double flanged beam. Sizing is the process of applying a protective coating on the warp yarns so that they can withstand repeated abrasion, stress, strain and flexing during the weaving process. Winding, warping and sizing are known as 'preparatory to weaving'. Then the warp yarns are drawn through the heald wires and reed dents in drawing and denting operations, respectively. Finally the fabric is manufactured on looms which perform several operations following a predefined sequence so that there is interlacement between warp and weft yarns. The general steps of woven fabric manufacturing are shown in Figure 1.3.

1.2.1 Types of Looms

Hand loom: This is mainly used in the unorganised sector. Operations such as shedding and picking are done by using manual power. This is one of the major sources of employment generation in rural areas of India and many other countries.

Power loom (non-automatic): All the operations of non-automatic power loom are driven by motor except pirn changing. They have very

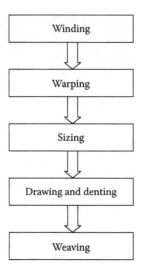

FIGURE 1.3   Steps of woven fabric manufacturing.

limited use for the production of normal woven fabrics, specially in modern industries. However, they are sometimes used for weaving industrial fabrics from very coarse weft (Marks and Robinson, 1976; Lord and Mohamed, 1982).

Automatic loom: In this power loom, the exhausted pirn is replenished by the full one without stoppage. This is possible only in under-pick system.

Multiphase loom: Multiple sheds can be formed simultaneously in this looms and thus productivity can be increased by a great extent. However, it has failed to attain commercial success.

Shuttleless loom: Weft is carried by projectiles, rapiers or fluids in case of shuttleless looms. The rate of fabric production is much higher for these looms. Besides, the quality of products is also better, and the product range is much broader compared to those of power looms. Most of the modern mills are equipped with different types of shuttleless looms based on the product range.

Circular loom: Tubular fabrics such as hosepipes and sacks are manufactured by circular looms.

Narrow loom: These looms are also known as needle looms and are used to manufacture narrow-width fabrics such as tapes, webbings, ribbons and zipper tapes.

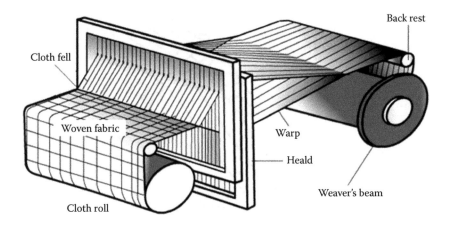

FIGURE 1.4 Basic loom components. (Reprinted from *Weaving: Conversion of Yarn to Fabric*, 2nd edn., Lord, P.R. and Mohamed, M.H, Copyright 1982, with permission from Elsevier.)

1.2.2 Primary Motions

Figure 1.4 shows some basic components of a loom. The longitudinal or warp yarns are supplied by weaver's beam positioned at the back of the loom. The produced fabric is wound over the cloth roller positioned at the front of the loom. For fabric manufacturing, three primary motions are required: shedding, picking and beat-up.

1.2.2.1 Shedding

Shedding is the process by which the warp sheet is divided into two groups so that a clear passage is created for the weft yarn or for the weft-carrying device to pass through it. One group of yarns either moves in the upward direction or stays in the up position (if they are already in that position) as shown in Figure 1.5, thus forming the top shed line. Another group of yarns either moves in the downward direction or stays in the down position (if they are already in that position), thus forming the bottom shed line.

Except for jacquard shedding, warp yarns are not controlled individually during the shedding operation. Healds are used to control a large number of warp yarns. The heald frame, which could be either metallic or wooden, carries a large number of metallic wires known as heald wires (Figure 1.6). Each heald wire has a hole, called heald eye, at the middle of its length. The warp yarn actually passes through the heald eye. Therefore, as the heald moves, all the warp yarns which are controlled by that head also move. The upward and downward movements of healds

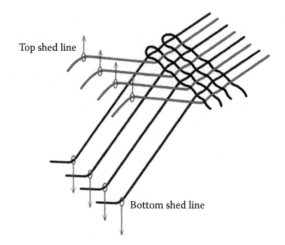

FIGURE 1.5   Shedding.

FIGURE 1.6   Heald.

are controlled either by cam or by dobby shedding mechanisms and associated heald reversing mechanism. The movement of the healds is not continuous. After reaching the topmost or lowest positions, the healds, in general, remain stationary for some duration. This is known as 'dwell'. In general, the shed changes after every pick, that is the insertion of weft.

1.2.2.2 Picking
The insertion of weft or weft-carrying device (shuttle, projectile or rapier) through the shed is known as picking. Based on the picking system, looms can be classified as follows.

- Shuttle loom: Weft package or pirn is carried by the wooden shuttle.

- Projectile loom: Weft is carried by metallic or composite projectile.

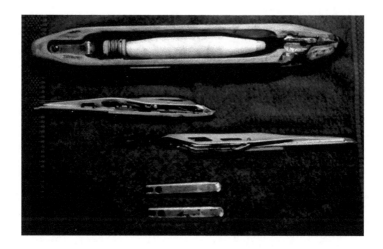

FIGURE 1.7   Shuttle, rapier heads and projectile (from top to bottom).

- Airjet loom: Weft is inserted by jet of compressed air.

- Waterjet loom: Weft is inserted by water jet.

- Rapier loom: Weft is inserted by flexible or rigid rapiers.

Figure 1.7 shows some weft-carrying devices.

With the exception of shuttle loom, weft is always inserted from only one side of the loom. The timing of picking is extremely important, especially in case of shuttle loom. The shuttle should enter the shed and leave the shed when the shed is sufficiently open (Figure 1.8). Otherwise, the movement of

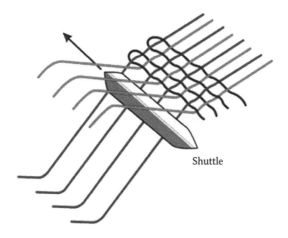

Shuttle

FIGURE 1.8   Picking.

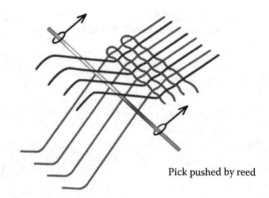

Pick pushed by reed

FIGURE 1.9   Beat-up.

the shuttle will be obstructed by the warp yarns, because of which the warp yarns may break or the shuttle may get trapped inside the shed, which may cause damage to reed, shuttle and warp yarns.

1.2.2.3 Beat-Up

Beat-up is the action by which the newly inserted weft yarn or pick is pushed up to the cloth fell (Figure 1.9). Cloth fell is the boundary up to which the fabric has been woven. The loom component responsible for the beat-up is called 'reed'. The reed, which is like a metallic comb, can have different count. For example, 80s Stockport reed has 80 dents in 2 inches. Generally, one or two warp yarns are passed through a single dent, and these are called 'one in a dent' or 'two in a dent', respectively. Reed is carried by the sley, which sways forward and backward due to the crank-connecting rod mechanism. This system is known as crank beat-up. In shuttleless looms, beat-up is done by cam mechanism, which is known as cam beat-up. Generally, one beat-up is done after the insertion of one pick.

1.2.3 Secondary Motions

For uninterrupted manufacturing of fabrics, two secondary motions are required. These are take-up and let-off. Take-up motion winds the newly formed fabric on the cloth roller either continuously or intermittently after the beat-up. The take-up speed also determines the picks/cm value in the fabric at loom state. As the take-up motion winds the newly formed fabric, tension in the warp sheet increases. To compensate this, the weaver's beam is rotated by the let-off mechanism so that adequate length of warp is released.

1.2.4 Auxiliary Motions

Auxiliary motions are mainly related to the activation of stop motions in case of any malfunctioning such as warp breakage, weft breakage or shuttle trapping within the shed. The major auxiliary motions are as follows:

- Warp stop motion (in case of warp breakage)

- Weft stop motion (in case of weft breakage)

- Warp protector motion (in case of shuttle trapping inside the shed)

1.2.5 Some Basic Definitions

1.2.5.1 Yarn Count

Yarn count represents the coarseness or fineness of yarns. There are two distinct principles to express the yarn count.

1. Direct systems (example: Tex, Denier)

2. Indirect systems (example: new English, i.e. Ne, Metric, i.e. Nm)

Direct systems revolve around expressing the mass of yarn per unit length. In contrast, indirect system expresses the length of yarn per unit mass. For example, 10 tex yarn implies that a piece of 1000 m long yarn will have a mass of 10 g. Similarly, for 10 denier, a piece of 9000 m long yarn will have a mass of 10 g. Denier is popularly used to express the fineness of synthetic fibres and filaments. A 10 denier yarn is finer than a 10 tex yarn as for the same mass, the length is nine times for the former.

On the other hand, 10 Ne implies that a 1-pound yarn will have a length of 10 × 840 yards. As the Ne value increases (say from 10 to 20 Ne), the yarn becomes finer.

Table 1.2 shows some popular yarn count systems.

TABLE 1.2   Direct and Indirect Systems of Yarn Count

Type	Name	Unit of Mass	Unit of Length
Direct	Tex	Gram	1000 m
	Denier	Gram	9000 m
Indirect	Ne	Pound	Hank (840 yards)
	Metric	Kilogram	Kilometre

The following conversion formula is used to change the yarn count from one system to another.

$$\text{Tex} = \frac{590.5}{\text{Ne}}, \quad \text{Denier} = 9 \times \text{Tex} \quad \text{and} \quad \text{Denier} = \frac{5315}{\text{Ne}}$$

1.2.5.2 Packing Factor or Packing Coefficient

Packing factor or packing coefficient represents the extent of closeness of fibres within the yarn structure. For the same yarn linear density, if the fibres are closely packed, then yarn diameter will be less. This happens when a spun yarn is manufactured with high level of twist. Figure 1.10 shows the close packing of circular fibres in a yarn.

Packing factor is expressed as follows:

$$\text{Packing factor} = \frac{\text{Cumulative area of all fibres}}{\text{Area of yarn cross section}} = \frac{\text{Yarn density}}{\text{Fibre density}}$$

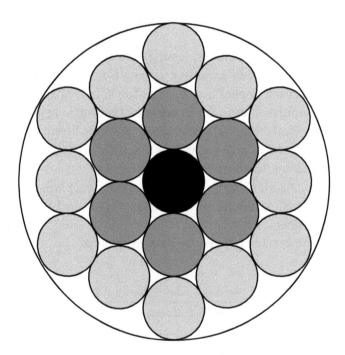

FIGURE 1.10   Close packing of circular fibres.

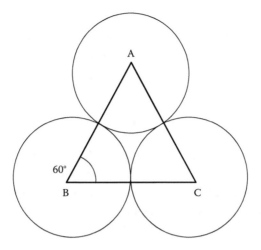

FIGURE 1.11   Repeat unit of close packing.

Figure 1.11 depicts the repeat unit of closely packed circular fibres. The equilateral triangle ABC actually indicates the repeat area. If fibre radius is r_f, then each side of the triangle ABC is having a length of $2r_f$.

Therefore,

$$\text{The area of triangle ABC} = \frac{\sqrt{3}}{4}\left(2r_f\right)^2 = \sqrt{3}r_f^2$$

$$\text{The total area of fibre inside the triangle ABC} = 3 \times \frac{1}{6}\pi r_f^2 = \frac{\pi}{2}r_f^2$$

So maximum possible packing factor

$$= \frac{\text{Total area of fibre inside triangle ABC}}{\text{Area of triangle ABC}} = \frac{(\pi/2)r_f^2}{\sqrt{3}r_f^2} = 0.91$$

For spun yarns, packing factor generally lies between 0.55 and 0.65. Yarns with lower packing factor are expected to be bulkier and softer. They can cause higher fabric cover for same fabric construction parameters.

1.2.5.3 Warp and Weft

A group of longitudinal yarns in a woven fabric (or on a loom) is called warp. A single warp is called 'end'. A group of transverse yarns in a woven fabric is called weft. A single weft is called 'pick'.

1.2.5.4 Crimp

Once the warp and weft are interlaced, both of them assume wavy or sinusoidal-like path. Thus the length of the yarn becomes more than that of the fabric within which the former is constrained. Crimp is a measure of the degree of waviness present in the yarns inside a woven fabric. Contraction is another measure of yarn waviness. The expressions of crimp and contraction are given below.

$$\text{Crimp \%} = \frac{\text{Length of yarn} - \text{Length of fabric}}{\text{Length of fabric}} \times 100 \qquad (1.1)$$

$$\text{Contraction \%} = \frac{\text{Length of yarn} - \text{Length of fabric}}{\text{Length of yarn}} \times 100 \qquad (1.2)$$

If the warp crimp is 10%, then the straightened length of an end, unravelled from the 1 m long fabric, will be 1.1 m.

1.2.5.5 Fractional Cover and Cover Factor

Fractional cover is the ratio of the area covered by the yarns to the total area of the fabric. If diameter of warp yarn is d_1 inch and spacing, that is gap between the two consecutive ends is p_1 inch, then fractional cover for warp (k_1) is d_1/p_1.

Now, for cotton yarns, having packing factor of 0.6, the relationship between yarn diameter (d) in inch and yarn count (Ne) is as follows:

$$d = \frac{1}{28\sqrt{\text{Ne}}} \qquad (1.3)$$

The relationship between end spacing (p_1) and ends per inch (n_1) is as follows:

$$n_1 = \frac{1}{p_1} \qquad (1.4)$$

After rearranging, the following expression is obtained for fractional cover for warp.

$$k_1 = \frac{n_1}{28\sqrt{Ne_1}} \tag{1.5}$$

where Ne_1 is the warp count.

Similarly, the expression for fractional cover for weft (k_2) is as follows:

$$k_2 = \frac{n_2}{28\sqrt{Ne_2}} \tag{1.6}$$

where

n_2 is the picks per inch

Ne_2 is the weft count

Cover factor is obtained by multiplying fractional cover with 28.

$$\text{Warp cover factor} = k_w = 28 \times k_1 = \frac{n_1}{\sqrt{Ne_1}} \tag{1.7}$$

Fabric cover is a very important parameter as it influences the following properties of woven fabrics:

- Air permeability

- Moisture vapour permeability

- Ultraviolet or any other types of radiation protection

Figure 1.12 shows two fabrics with low and high cover factors.

1.2.5.6 Porosity

Porosity is a measure of presence of void or air inside the fabric or fibrous assemblies. It indicates the percentage of volume of fabric that has been occupied by the air. If there is no porosity, then the densities of the fibre and fabric will be the same. However, in most of the fabrics there will be some air pockets which will contribute to the volume and not to the mass. For example, woven and knitted fabrics can have typical porosity of 70%–80% and 80%–90%, respectively.

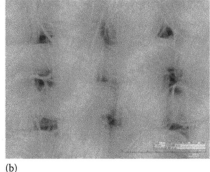

(a)   (b)

FIGURE 1.12   Fabrics with (a) low and (b) high cover factors.

Let us consider the following:

Fabric areal density or gram per square meter (GSM) = G g/m²

Thickness of fabric = T m

Density of fibre = ρ g/m³

Porosity (%) = P

$$\text{The mass of 1 m}^3 \text{ fabric will be} = \frac{\text{Areal density}}{\text{Thickness}} = \frac{G}{T} \text{ g}$$

Since the mass of fabric is only being contributed by the fibre, the mass of 1 m³ fabric will be = Volume occupied by the fibre in 1 m³ fabric × density of fibre = $\left(1 - \dfrac{P}{100}\right) \times \rho$ g.

$$\text{Therefore,} \left(1 - \frac{P}{1000}\right) \times \rho = \frac{G}{T}$$

$$\text{So porosity}(\%) = \left(1 - \frac{G}{T \times \rho}\right) \times 100 \qquad (1.8)$$

Porosity influences the thermal conductivity of the fabric or fibrous assemblies. Air is a poor conductor of heat, and its thermal conductivity (K_{air}) is 0.025 W/m K. On the other hand, thermal conductivity of cotton fibre is around 0.24 W/m K, which is approximately 10 times more than that of air. The thermal conductivity of a fabric having porosity of P can be expressed as follows:

$$\text{Thermal conductivity of fabric} = \left(1 - \frac{P}{100}\right)K_{fibre} + \frac{P}{100} \times K_{air} \quad (1.9)$$

Therefore, higher porosity implies lower thermal conductivity of the fabric and vice versa.

1.2.5.7 Areal Density

Areal density is expressed by the mass of the fabric per unit area. In most of the cases, the mass is expressed in gram (g) and area is expressed in square meter (m²). Therefore, the unit becomes g/m², which is popularly called GSM. Areal density of the fabric will depend on the following parameters:

- Warp yarn count (tex): T_1
- Weft yarn count (tex): T_2
- Ends per unit length (EPcm): N_1
- Picks per unit length (PPcm): N_2
- Crimp % in warp: C_1
- Crimp % in weft: C_2

Let us consider a piece of fabric having dimensions of 1 m × 1 m as shown in Figure 1.13. The number of ends per cm is N_1. So, the total number of ends in the given fabric is $100N_1$. The projected length of one end is 1 m when incorporated in the fabric. However, the end has some crimp in it. Therefore,

$$\text{Straightened length of one end} = 1 \times \left(1 + \frac{C_1}{100}\right) \text{ m}$$

$$\text{Total length of ends (warp)}$$

$$= \text{Total number of ends} \times \text{straightened length of one end}$$

$$= 100N_1 \times \left(1 + \frac{C_1}{100}\right) \text{ m.}$$

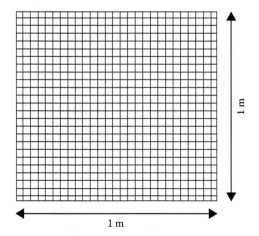

FIGURE 1.13   One square meter of fabric.

$$\text{Mass of warp yarns (g)} = \frac{\text{Total length of ends in m}}{1000} \times \text{tex of warp}$$

$$= \frac{100 N_1 \times \left(1 + \dfrac{C_1}{100}\right)}{1000} \times T_1 = \frac{N_1 T_1}{10}\left(1 + \frac{C_1}{100}\right)$$

Similarly

$$\text{The mass of weft yarns (g)} = \frac{N_2 T_2}{10}\left(1 + \frac{C_2}{100}\right)$$

So

Total mass of the fabric (g) = Mass of warp yarns (g) + Mass of weft yarns (g)

$$= \frac{N_1 T_1}{10}\left(1 + \frac{C_1}{100}\right) + \frac{N_2 T_2}{10}\left(1 + \frac{C_2}{100}\right)$$

$$= \frac{1}{10}\left[N_1 T_1 \left(1 + \frac{C_1}{100}\right) + N_2 T_2 \left(1 + \frac{C_2}{100}\right) \right]$$

So

$$\text{Areal density of fabric or GSM} = \frac{1}{10}\left[N_1 T_1 \left(1 + \frac{C_1}{100}\right) + N_2 T_2 \left(1 + \frac{C_2}{100}\right) \right]$$

$$(1.10)$$

TABLE 1.3   Cover, Areal Density, Thickness and Porosity Values of Fabrics

Fabric Type	Cover (%)	Areal Density (g/m^2)	Thickness (mm)	Porosity (%)
Polyester-cotton	78.4	122	0.298	72.9
Polyester-cotton	84.6	136	0.307	70.7
100% cotton	88.2	135	0.402	77.8
100% cotton	92.9	155	0.424	75.8
100% cotton	95.2	171	0.442	74.4

Table 1.3 presents cover, areal density, thickness and porosity values of some woven fabrics made from spun yarns.

1.3 KNITTING TECHNOLOGY

Knitting is a process of fabric formation by producing series of intermeshed loops. Loops are the building blocks of knitted fabrics (Figure 1.14). The upper part of the loop is called 'head', whereas the two sides are called 'legs'. The intermeshing of two loops happens through the 'foot'. In general, the knitted fabrics are more stretchable than the woven fabrics. The open structure of knitted fabrics also facilitates better moisture vapour transmission, making it suitable for sports garments and high-activity clothing. Besides, the knitted fabrics have more porosity than the woven fabrics. Therefore, knitted fabrics

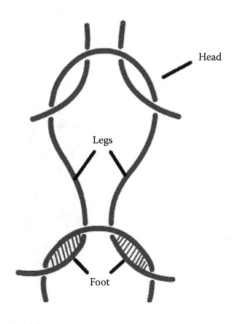

FIGURE 1.14   A knitted loop.

can trap more air, resulting in lower thermal conductivity and higher thermal resistance. There are two types of knitting: warp knitting and weft knitting.

1.3.1 Weft Knitting

In weft knitting loops are made by the supplied yarns across the width of the fabric (Figure 1.15a). Weft knitted fabric can be made even from one supply package. The weft knitting machines are of two types:

1. Flatbed machine

   a. Single bed

   b. Double bed or V bed

2. Circular bed machine

In flatbed machine, the needles do not perform any lateral movement. The axial movement of the needles, needed for loop formation, is actuated by a set of cams mounted on cam jacket which reciprocate laterally (exception: straight bar machines). In contrast, the cam jackets are generally stationary in circular knitting machine, whereas the cylinder carrying the needles on its grooved surface rotates continuously to cause the upward and downward movement of needles. In many small-diameter circular weft knitting machines, the cylinder may remain stationary while the cam jackets revolve. This is true for single feeder machines.

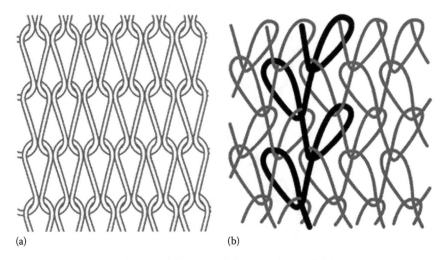

(a)   (b)

FIGURE 1.15   (a) Weft knitted fabric and (b) warp knitted fabric.

1.3.2 Warp Knitting

In warp knitting, loops are made from each warp yarn along the length of the fabric (Figure 1.15b). The yarns are supplied in the form of a sheet made by parallel warp yarns coming out from a single or multiple warp beams. The yarns are fed to the needles by guide bars which swing to and fro and shog laterally. The loop formation mechanism is more complex for warp knitting than that of weft knitting.

1.3.3 Needle

Irrespective of the knitting technology, the machine element which helps in loop formation is called needle. Latch, bearded and compound needles are used, depending on the type of knitting machine. Latch needle (Figure 1.16) is most popularly used in weft knitting and Raschel warp knitting machine.

The major components of a latch needle are as follows:

- Hook

- Latch

- Latch spoon (cup)

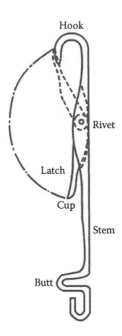

FIGURE 1.16   Latch needle.

- Stem

- Butt

Hook is the curved part of the needle which is responsible for loop formation. Latch is a tiny component that is riveted on the upper part of stem of the needle. Latch spoon (cup) is the tip of the latch which encloses the outer surface of the tip of the hook when the latch closes. The upward and downward movements of the needles during the loop formation are caused by a set of cams in weft knitting and by movement of bars in warp knitting. The butt is actually the 'follower', and it is pressed against the cam to cause movement of the needle.

1.3.4 Loop Formation in Knitting

The sequence of loop formation is shown in Figure 1.17. When the needle moves up, the old loop forces the latch to open. When the old loop rests on the latch, the position is called 'tuck' position (1 in Figure 1.17). Then the needle moves up further and the old loop slides down the latch and rests on the stem of the needle. This is called the 'clearing' position (2 in Figure 1.17). The needle attains its highest position at 3 in Figure 1.17. Then it starts to descend and the hook catches the yarn. As the needle continues to descend, the yarn bends in the form of a loop ('U' shape). The old loop now helps close the latch by pushing it in upward direction so that the newly formed loop is about to be caught between the hook and latch (4 in Figure 1.17). The needle

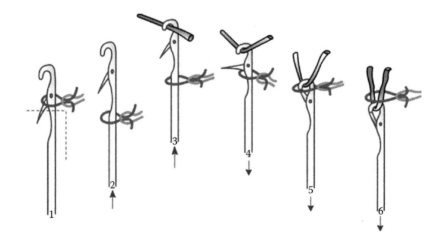

FIGURE 1.17   Sequence of loop formation.

continues to descend and new loop is 'cast on' (5 in Figure 1.17) and finally 'knocked over' (6 in Figure 1.17) through the old one. Casting-off or knocking-over is the same phenomenon, executed in two different manners. For casting-off to take place, special knitting elements bodily push the old loop out, while in case of knocking-over, help of sinkers or verges is necessary to prevent the old loop from moving down with the needle.

1.3.5 Course and Wale

The horizontal row of loops is called 'course'. The vertical column of loops is called 'wale' (Figure 1.18). The wales per inch (*wpi*) and courses per inch (*cpi*) of knitted fabrics are analogous to ends per inch (*epi*) and picks per inch (*ppi*) of woven fabrics. For a fully relaxed knitted fabric, the *wpi* and *cpi* values are determined by the loop length. Smaller loop length leads to higher values of *wpi* and *cpi*. As a result, the stitch density or loop density, which is a product of *wpi* and *cpi*, also increases with the reduction in loop length. The ratio of *cpi* and *wpi* is known as loop shape factor. For a fully relaxed single jersey fabric, the loop shape factor is around 1.3.

Let the loop length be *l* inch.

So

$$\text{Wales per inch}\left(wpi\right) = \frac{k_w}{l} \quad \text{and} \quad \text{courses per inch}\left(cpi\right) = \frac{k_c}{l} \quad (1.11)$$

FIGURE 1.18   Course and wale.

Therefore,

$$\text{Stitch density (inch}^{-2}) = wpi \times cpi = \frac{k_w}{l} \times \frac{k_c}{l} = \frac{k_w k_c}{l^2} = \frac{k_s}{l^2} \quad (1.12)$$

where k_w, k_c and k_s are wale constant, course constant and stitch constant, respectively. These constants are independent of yarn and machine variables. In a fully relaxed state, the values of these constants for worsted single jersey fabric are 4.2, 5.5 and 23.1, respectively.

1.3.6 Single Jersey and Double Jersey Fabrics

Flatbed machines, as the name implies, have one or more beds for carrying the needles. Single bed machines produce plain or single jersey structure, whereas double bed machines produce rib (1×1, 2×2 etc.) and purl structures. Single jersey and double jersey (1×1 rib) structures are shown in Figure 1.19. Double bed machines can also be employed to develop single jersey constructions. The needles on the two beds in a double bed machine must be offset so that they do not collide with each other while forming the loops.

In case of single jersey fabrics, all the heads of the loops either face the viewer or are away from the viewer. In Figure 1.19a, all the heads of the loops are hidden from the viewer while the feet are prominently visible, so it is the technical face side of the fabric (Banerjee, 2015). The other side of the fabric is known as technical back.

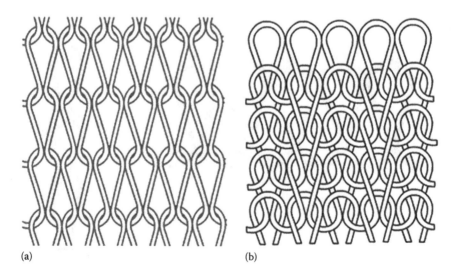

(a)   (b)

FIGURE 1.19 (a) Single jersey and (b) double jersey (rib) structures.

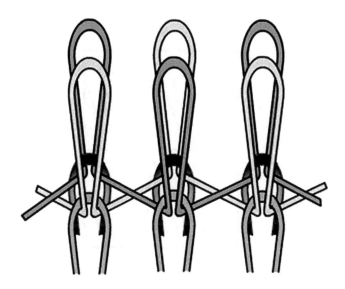

FIGURE 1.20   Interlock structure.

In case of rib (double jersey) fabrics, in some of the wales, heads of the loops face the viewer and vice versa (Figure 1.19b), so there is no technical face or back in double jersey fabric. Single jersey fabrics tend to curl at the edges. In general, double jersey fabrics are thicker and more stretchable in course direction than the single jersey fabrics.

In circular single jersey knitting machine, only one set of needles are used on the cylinder. However, in circular rib knitting machine, two sets of needles – cylinder and dial needles – are used. They operate perpendicularly to each other. One set of needles (cylinder needles) are arranged on the surface of a grooved cylinder. Generally the cylinder is rotated and needles get the requisite movement from the stationary cam jackets. Another set of needles operate in horizontal plane and they are known as dial needles.

Another important double jersey structure which is made on circular machines is known as interlock. It is basically a combination of two rib structures as shown in Figure 1.20. The interlocking of two rib structures is responsible for lower stretchability of interlock fabrics as compared to the original rib structure. Interlock fabrics are generally heavy and demonstrate least porosity among the three knitted structures (single jersey, rib and interlock).

1.3.7 Tightness Factor

Tightness factor implies the relative tightness or looseness of a single jersey knitted fabric by indicating the ratio of fabric area covered by the yarn

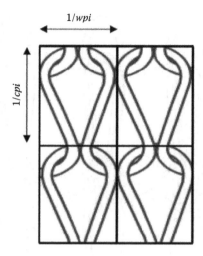

FIGURE 1.21 Repeat unit of a knitted loop.

to the total fabric area corresponding to one loop. It is analogous to the cover factor of woven fabric.

Figure 1.21 depicts the repeat unit of a single jersey knitted fabric. The total area of the repeat (in square inch) is equal to the product of $1/cpi$ and $1/wpi$. The area covered by the yarn can be calculated by the product of loop length (l inch) and yarn diameter (d inch), assuming that the loop has a planner structure.

Therefore,

$$\text{Tightness factor} = \frac{\text{Area covered by the yarn within the repeat unit}}{\text{Area of the repeat unit}}$$

$$= \frac{l \times d}{\dfrac{1}{cpi} \times \dfrac{1}{wpi}} = \frac{l \times d}{\dfrac{l}{k_c} \times \dfrac{l}{k_w}} = \frac{k_c \times k_w \times d}{l} = \frac{k_s \times d}{l}$$

Yarn diameter (d) is proportionate with $(\text{tex})^{0.5}$.

Therefore,

$$\text{Tightness factor} \propto \frac{k_s \times \sqrt{\text{tex}}}{l} \tag{1.13}$$

When the knitted structure is same and the relaxation state of the fabric is also same, then $\sqrt{\text{tex}}/l$ can be used for comparison of tightness factor (Ray, 2012).

1.4 NONWOVEN TECHNOLOGY

A nonwoven is a sheet of fibres, continuous filaments or chopped yarns of any nature or origin that have been formed into a web by any means and bonded together by any means, with the exception of weaving or knitting. Felts obtained by wet milling are not nonwovens (www.edana.org).

Nonwovens are engineered and flat structured sheets which are made by bonding and entangling of fibres by means of mechanical, thermal or chemical processes. Nonwoven technology has attracted the attention of researchers and industrialists as it can manufacture the fabric at a very high production rate bypassing the yarn production stage. The principal end-uses of nonwoven materials are found in technical textiles such as geotextiles, filtration, wipes, health and hygiene products, surgical gowns, face masks, automotive textiles and so on. The two major stages of nonwoven manufacturing are web formation and web bonding. The major nonwoven technologies can be listed as follows:

Web Formation	Web Bonding
Mechanically formed fibre webs (Drylaid)	Needle punching
Aerodynamically formed fibre webs (Drylaid)	Hydro-entanglement
Hydrodynamically formed fibre webs (Wetlaid)	Thermal bonding
Polymer-laid (Spunmelt nonwovens)	Chemical bonding

Drylaid, wetlaid and polymer-laid web formation systems have their roots in textile, paper making and polymer extrusion processes, respectively.

1.4.1 Needle Punching Technology

Needle punching is the most common web bonding method. It is the method of consolidation of fibrous webs by repeated insertion of barbed needles into the web as shown in Figure 1.22. The needling can be done either from one side or from both (top and bottom) sides of the web. This process consolidates the structure of fibrous web by interlocking of fibres in the third or 'Z' dimension without using any binder. Continuous filaments or short staple fibres are initially arranged in the form of a fibrous web in various orientations (random, cross, parallel or composite).

The degree of compaction of the fibrous web is largely dependent on punch density (*PD*), which is defined as the number of needle penetrations received by the fibrous web per unit area (punches/cm²). If the stroke frequency (cycles/min) of the needle board is N and the number of needles per

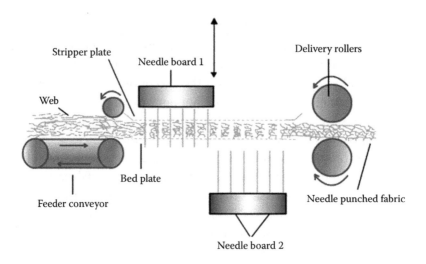

FIGURE 1.22   Needle punching process.

linear cm width of the needle board is *n* and the delivery speed of the fabric is *v* cm/min, then punch density can be calculated by the following expression.

$$\text{Punch density } PD = \frac{N \times n}{v} \text{ cm}^{-2} \tag{1.14}$$

Therefore, increase in stroke frequency or the number of needles per cm width of the needle board will increase the punch density for a given delivery speed. Higher punch density creates more fibre entanglement in the web, thus increasing the compactness of the web.

Needle-punched nonwoven geotextiles are extensively used in civil engineering applications, including road and railway construction, landfills, land reclamation and slope stabilisation. Such applications require geotextiles to perform more than one function, including filtration, drainage and separation. The properties of needle-punched nonwoven fabric depend on parameters such as fibre type, web aerial density, needle penetration depth, punch density (number of punches/cm²) and the number of needling passages.

1.4.2 Hydro-Entanglement Technology

Hydro-entanglement, or spunlacing, is a versatile method of bonding the fibrous web using high-pressure water jets. In this process, a fabric is produced by subjecting a web of loose fibres to high-pressure fine water jets as shown in Figure 1.23. The fibre web is supported by either regularly spaced woven wires or a surface with randomly distributed

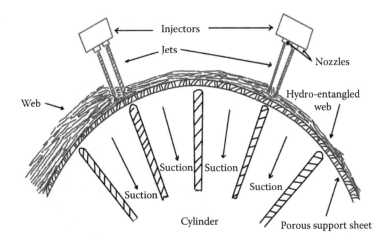

FIGURE 1.23 Hydro-entanglement process. (From Banerjee, P.K., *Principles of Fabric Formation*, CRC Press, Boca Raton, FL, 2015.)

holes (Perfojet technology). As a result of the impact of high-pressure jets, the fibres bend and curl around each other, forming an integrated web where fibres are held together by frictional forces. The fabrics are finally dried to remove the water.

The resulting fabric strength depends on the web properties (areal density, thickness, etc.), fibre parameters (fibre diameter, cross-section, bending modulus, etc.) and forming wires geometry and jet parameters. Hydro-entangled nonwovens have an extensive range of applications, including wipes, carpet backing, filters, sanitary, medical dressings and composites. Among these applications, personal care and household wipes form the fastest growing sector. Use of natural resources like water and energy (for drying) is a concern with this technology.

1.4.3 Spunbond Technology

Spunbond and meltblown fabrics belong to the class of polymer-laid nonwovens. In spunbonding process, fluid polymer is converted into finished fabric by a series of continuous operations as shown in Figure 1.24. Polymer melt is first extruded into filaments and then the filaments are attenuated. While the filaments are being attenuated, they remain under tension. After attenuation, the tension is released and the filaments are forwarded to a surface where the web is formed. The web is then subjected to the bonding process which can be done by chemical and/or thermal process. A binder may be incorporated in the spinning process or applied subsequently.

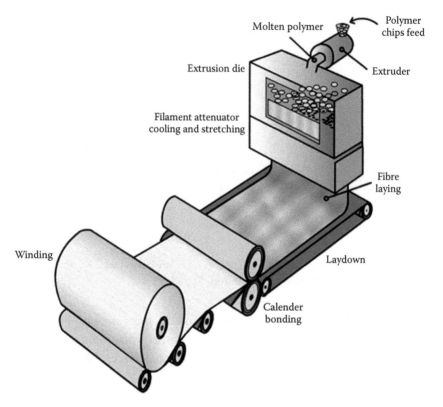

FIGURE 1.24   Spunbonding process.

Polypropylene and polyester are commonly used for spun-bonding process. Spun-bonded nonwovens have high strength but lower flexibility.

1.4.4 Meltblown Technology

Meltblown technology is unique in the sense that the web is very fine along with very small pore sizes which no other nonwoven technology can match. The polymer is fed into the die tip and the resulting fibre is attenuated by hot air, which is blown near the die tip as shown in Figure 1.25. Air and fibre are expanded into the free air. Due to the mixture of high-speed air and fibre with ambient air, the fibre bundle starts its movement forward and backward creating 'form drag'. This form drag appears with every change of fibre direction. Therefore the meltblown fibres usually do not have a constant diameter. The fibres in the meltblown web are laid together by a combination of entanglement and cohesive sticking. The ability to form a web directly from a molten polymer without controlled stretching gives meltblown technology a distinct cost advantage over other systems.

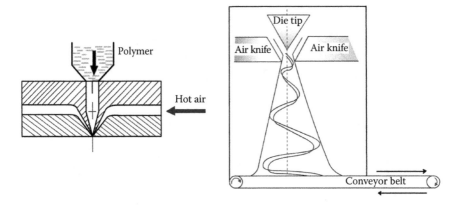

FIGURE 1.25   Schematic diagram of meltblown process.

Meltblown webs offer a wide range of characteristics such as random fibre orientation, very fine fibre and low to moderate web strength. The major end-uses of monolithic meltblown fabrics are filtration media (surgical mask filters, liquid and gas filtration), surgical disposable gown, sterilisation warp, disposable absorbent products and oil absorbents. About 40% of meltblown fabrics are used in uncombined (monolithic) state. The laminated SMS (spunbond-meltblown-spunbond) structures are ideal for gradient filtration as the material shows excellent barrier properties combined with mechanical strength. The filtration efficiency is often increased by applying electrostatic charges to the filaments. With this standard meltblown technology, fibres of 1–3 μm diameter can be produced. Several filter applications using meltblown nonwovens are already in the market today.

1.5 BRAIDING TECHNOLOGY

Braiding generally produces tubular or narrow fabrics by intertwining three or more strands of yarns, threads or filaments. The yarn packages move on serpentine path as shown in Figure 1.26. In simple machines, half of the packages move in clockwise direction, whereas the remaining packages move in anticlockwise direction. Shoelaces and ropes are manufactured using braiding systems. Profiled braided structures are also used as composite performs. The interlacement pattern of braided structures has resemblance with that of woven structures. For example, Diamond, Regular and Hercules braids have interlacement patterns similar to those of plain, 2 × 2 and 3 × 3 twill weaves, respectively.

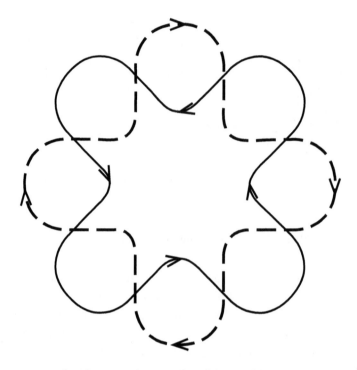

FIGURE 1.26   Paths of yarn packages in braiding machine.

During the braiding process, the yarns form helical paths around a mandrel. The braid angle is defined as the angle between the yarn axis and the braid axis. This is similar to the twist angle of a fibre inside the yarn. The braid angle is a very important parameter for the braided structure and can be calculated using the following expression.

$$\text{Braid angle} = \theta = \tan^{-1}\left(\frac{\omega r}{TUS}\right) \tag{1.15}$$

where
   ω is the average angular velocity of the package (rad/s)
   r is the mandrel radius (cm)
   TUS is the take-up speed (cm/s)

If the number of yarn careers or yarn packages is K, then the braided structure forms $K/2$ number of parallelograms in the circumferential direction. This is shown in Figure 1.27.

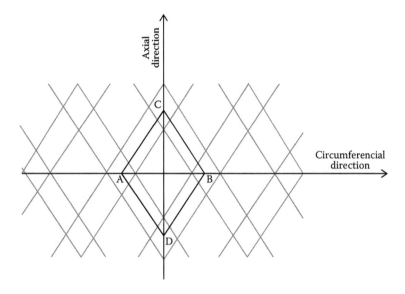

FIGURE 1.27   Unit cell of braided structure.

The cover factor for the above braided structure can be calculated using the following expression (Potluri et al., 2003).

$$\text{Cover factor} = 1 - \left(1 - \frac{W_y K}{4\pi r \cos\theta}\right)^2 \tag{1.16}$$

where
   W_y is the yarn width
   r is the mandrel radius
   θ is the braid angle

NUMERICAL PROBLEMS

1.1   The length of a fabric is 10 m. The length of a warp yarn, removed from the fabric, in straight condition is 10.8 m. Determine the crimp % and contraction % in warp direction.

   Solution:

$$\text{Crimp \%} = \frac{L_{yarn} - L_{fabric}}{L_{fabric}} \times 100$$

   where
      L_{yarn} is the length of warp yarn removed from the fabric
      L_{fabric} is the length of fabric

So

$$\text{Crimp} = \frac{(1.08 - 1)}{1} \times 100 = 8\%$$

$$\text{Contraction} = \frac{L_{yarn} - L_{fabric}}{L_{yarn}} \times 100$$

$$= \frac{1.08 - 1}{1.08} \times 100 = 7.41\%$$

So, the crimp and contraction are 8% and 7.41%, respectively.

1.2 Prove that for cotton yarn with packing factor of 0.6, diameter $(\text{inch}) = \dfrac{1}{28\sqrt{\text{Ne}}}$

Solution:

Figure 1.28 shows a cylindrical yarn. The count of the yarn is Ne. The diameter and length of the yarn (for mass of 1 pound) is d inch and l inch, respectively.

As the yarn count is 'Ne', there will be Ne number of hanks (840 yards) in 1 pound.

Thus length (l) = Ne × 840 × 36 inch.

Density of cotton fibre is 1.51 g/cm³.

As the packing factor is 0.6, the density of the cotton yarn will be 1.51 × 0.6 = 0.906 g/cm³.

So

$$\text{The density of cotton yarn} = \frac{0.906 \times 2.54^{3}}{453.6} \text{ pound/inch}^{3}$$

$$= 0.0327 \text{ pound/inch}^{3}$$

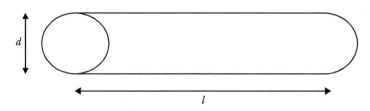

FIGURE 1.28 Yarn with circular cross section.

$$\text{The volume of yarn} = \frac{\pi d^2}{4} l \text{ inch}^3$$

$$\text{Mass of the yarn} = \frac{\pi d^2}{4} l \times 0.0325 \text{ pound}$$

$$= \frac{\pi d^2}{4} \times \text{Ne} \times 840 \times 36 \times 0.0327 \text{ pound}$$

$$= 1 \text{ pound (by definition)}$$

Therefore,

$$d\,(\text{inch}) = \frac{1}{28\sqrt{\text{Ne}}}$$

1.3 A cotton fabric is made from 20 Ne warp and ends per inch is 50. Determine the warp cover factor.

Solution:

Two consecutive ends are shown in Figure 1.29. Here, ends per inch = 50

$$\text{End spacing}\,(p) = \frac{1}{50''} = 0.02 \text{ inch}$$

Warp count = 20 Ne
So

$$\text{Warp yarn diameter}\,(d) = \frac{1}{28\sqrt{\text{Ne}}} \text{ inch}$$

$$= \frac{1}{28\sqrt{20}} = 0.008 \text{ inch}$$

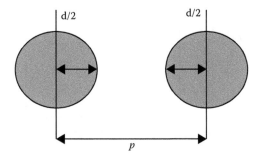

FIGURE 1.29 Spacing between two ends.

Now,

$$\text{Warp cover factor} = 28 \times \frac{d}{p}$$

$$= 28 \times \frac{0.008}{0.02} = 11.2$$

So warp cover factor is 11.2.

1.4 Show that the expression for fabric cover factor is $K_1 + K_2 - \dfrac{K_1 K_2}{28}$, where K_1 is warp cover factor and K_2 is weft cover factor.

Solution:
Figure 1.30 shows the repeat unit of a plain woven fabric.
Let d_1 is the diameter of warp yarn,
d_2 is the diameter of weft yarn,
p_1 is the end spacing,
p_2 is the pick spacing.

$$\text{Fractional cover} = \frac{\text{Area covered by the yarns within the repeat}}{\text{Area of the repeat}}$$

$$= \frac{d_1 p_2 + d_2 p_1 - d_1 d_2}{p_1 p_2}$$

$$= \frac{d_1}{p_1} + \frac{d_2}{p_2} - \frac{d_1 d_2}{p_1 p_2}$$

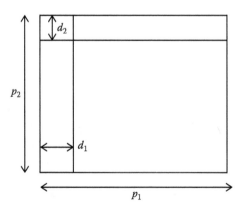

FIGURE 1.30 Repeat unit of a plain woven fabric.

$$\text{Cover factor} = 28 \times \text{Fractional cover}$$

$$= 28\frac{d_1}{p_1} + 28\frac{d_2}{p_2} - 28\frac{d_1 d_2}{p_1 p_2}$$

So

$$\text{Fabric cover factor } \left(K_f\right) = K_1 + K_2 - \frac{K_1 K_2}{28}$$

1.5 Calculate the areal density (g/m²) of the cotton fabric having the following specifications: Warp and weft count: 22s × 18s; 25 ends per cm × 16 picks per cm; warp crimp is 6.5% and weft crimp is 8.5%.

Solution:

$$\text{Warp count is } 22 \text{ Ne} = \frac{590.5}{22}\text{tex} = 26.84 \text{ tex } \left(T_1\right)$$

$$\text{Weft count is } 18 \text{ Ne} = \frac{590.5}{18} = 32.81 \text{ tex } \left(T_2\right)$$

Ends per cm (N_1) = 25
Picks per cm (N_2) = 16
Warp crimp (C_1) = 6.5%
Weft crimp (C_2) = 8.5%

$$\text{Areal density of fabric} = \frac{1}{10}\left[N_1 T_1\left(1+\frac{C_1}{100}\right) + N_2 T_2\left(1+\frac{C_2}{100}\right)\right]$$

$$= \frac{1}{10}\left[25 \times 26.84\left(1+\frac{6.5}{100}\right) + 16 \times 32.81\left(1+\frac{8.5}{100}\right)\right]$$

$$= 128.42 \text{ g/m}^2$$

So, the areal density of the fabric is 128.42 g/m².

1.6 Calculate the porosity and thermal conductivity of a needle-punched nonwoven fabric made of 100% polypropylene fibres (density is 0.9 g/cm³) having the thickness of 3 mm. The areal density of the nonwoven fabric is 300 g/m² and the thermal conductivity of polypropylene fibre is 0.12 W/m K.

Solution:

Let fabric areal density or GSM is G g/m^2, thickness of fabric is T m and density of fibre is ρ g/m^3.

$$\text{Then porosity}(\%) = \left(1 - \frac{G}{T \times \rho}\right) \times 100$$

where

   $T = 3$ mm $= 0.003$ m

   $\rho = 0.9$ g/cm$^3 = 0.9 \times 10^6$ g/m^3

So

$$\text{Porosity} = \left(1 - \frac{300}{0.003 \times 0.9 \times 10^6}\right) \times 100$$

$$= (1 - 0.111) \times 100 = 88.89\%$$

$$\text{Thermal conductivity of fabric} = \left(1 - \frac{P}{100}\right) K_{fibre} + \frac{P}{100} \times K_{air}$$

$$= \left(1 - \frac{88.89}{100}\right) 0.12 + \frac{88.89}{100} \times 0.025$$

$$= 0.036 \text{ W/m K}$$

So the porosity and thermal conductivity are 88.89% and 0.036 W/m K, respectively.

REFERENCES

Banerjee, P. K. 2015. *Principles of Fabric Formation.* Boca Raton, FL: CRC Press.

Lord, P. R. and Mohamed, M. H. 1982. *Weaving: Conversion of Yarn to Fabric,* 2nd edn. Merrow, UK: Merrow Technical Library.

Marks, R. and Robinson, A. T. C. 1976. *Principles of Weaving.* Manchester, UK: The Textile Institute.

Potluri, P., Rawal, A., Rivaldi, M. and Porat, I. 2003. Geometrical modelling and control of a triaxial braiding machine for producing 3D performs. *Composites: Part A,* 34: 481–492.

Ray, S. 2012. *Fundamentals and Advances in Knitting Technology.* New Delhi, India: Woodhead Publishing India Pvt. Ltd.

CHAPTER 2

Winding

2.1 INTRODUCTION

Winding is the first preparatory process for weaving. Readers who have experienced kite flying would be able to visualise the similarities between the yarn preparation for kite flying and yarn preparation for weaving. The yarn to be used for kite flying is wound uniformly on a spool after applying a protective coating on the yarn. The protective coating is prepared by concocting a mixture of starch, glass powder and some other materials so that the abrasion resistance of the yarn improves. Similarly, in weaving preparatory, yarns are wound in a systematic manner on the package during winding process. Moreover, in sizing process, a protective coating is applied on the yarn to improve its abrasion resistance and weaving performance.

2.2 OBJECTIVES

- To warp the forming yarn on a package in a systematic manner or to transfer yarn from one supply package to another in such a way that the latter is adequately compact and usable for the subsequent operations.

- To remove the objectionable faults present in the original yarns.

Most of the textile winding operations deal with the conversion of ringframe bobbins into cones or cheeses. One ringframe bobbin (cop) typically contains around 100 g of yarn. If the yarn count is 20 tex, then the length of yarn in the package will be around 5 km. As the warping speed in modern machines is around 1000 m/min, direct use of ringframe bobbins in warping will necessitate supply package change after every 5 min. This will reduce the running efficiency of warping machine. Besides, there will be a large number of yarn breaks during warping as the yarn in ringframe bobbin contains many thin and weak places. Therefore, ringframe bobbins are converted into bigger cones (mass around 2 kg or more) or cheeses.

Ringframe bobbins are also not useable as weft packages because they have empty core which will require bigger size of the shuttle and thereby causing problem in shedding operation. Therefore, for shuttle looms, pirn winding is carried out to produce weft packages (pirn) from cones.

2.3 TYPES OF YARN WITHDRAWAL

Two basic motions are required for effective winding. First, the rotational motion of the package, on which the yarn is being wound, is required. This rotational motion pulls out the yarn from the supply package. Second, the traverse motion is requited so that the entire length of the package is used for winding the yarn. In the absence of the latter, yarn will be wound at the same region by placing one coil over another, which is not desirable.

During winding, the yarn can be withdrawn from the supply packages in two ways as depicted in Figure 2.1.

- Side withdrawal

- Over-end withdrawal

Side withdrawal is preferable for flanged packages as the yarn does not touch the flanges. The package has to rotate during the yarn withdrawal. However, for ringframe bobbins, over-end withdrawal is performed by keeping the package almost in upright condition. As one coil comes out from the ringframe bobbin, one twist is either added or subtracted from parent yarn depending on the direction of twist in the yarn. This can be visualized from Figure 2.2. A flat ribbon is wound on a ringframe bobbin or cop. When the ribbon is withdrawn over-end, the ribbon gets twisted. This implies that over-end withdrawal causes twist variation in yarn. If the twist per cm in original yarn is 8 and circumference of one coil is 10 cm, then the total number of twist in one coil is 80. After unwinding, one twist will be either added or subtracted. Therefore, the number of twist in 10 cm length of yarn will be 80 ± 1. So the twist per cm in yarn on the delivery package will be either 7.9 or 8.1.

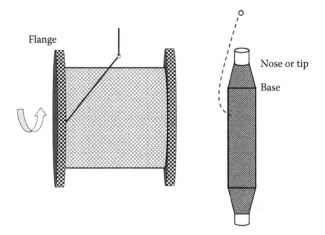

FIGURE 2.1   Side withdrawal (left) and over-end withdrawal (right).

FIGURE 2.2   Twist variation in over-end withdrawal.

2.4 TYPES OF WOUND PACKAGES

There could be three types of wound packages based on the angle at which the yarns are laid on the package.

1. Parallel wound package

2. Nearly parallel wound package

3. Cross wound package

Figure 2.3 depicts various types of wound packages.

In parallel wound package, yarns are laid parallel to each other. This helps to maximize the yarn content in the package. However, parallel wound packages suffer from the problem of stability and the layers of

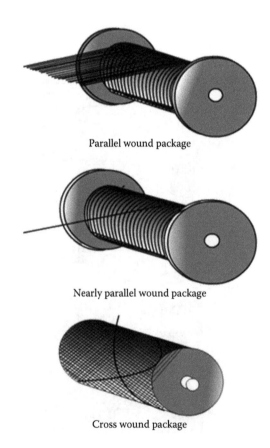

Parallel wound package

Nearly parallel wound package

Cross wound package

FIGURE 2.3   Various types of wound packages.

coils can collapse, especially from the two sides of the package. Therefore, double flanged packages are generally used for parallel wound packages.

Example: Weaver's beam, warper's beam.

In nearly parallel wound package, successive coils of yarn are laid with a very nominal angle between them. The rate of traverse is very slow in this case.

Example: Flanged bobbin, roving bobbin.

In cross wound package, successive coils of yarn are laid on the package at considerable angle between them. As the coils crosses each other very frequently, the package content and package density are lower than those of parallel wound package. However, cross wound package provides very good package stability as the coils often change their direction at the edges of the package.

Example: Cone, cheese.

2.5 IMPORTANT DEFINITIONS OF WINDING

2.5.1 Wind

It is the number of revolutions made by the package (i.e. the number of coils wound on the package) during the time taken by the yarn guide to make a traverse in one direction (say from left to right) across the package.

2.5.2 Traverse Ratio or Wind Ratio or Wind per Double Traverse

It is the number of revolutions made by the package (i.e. the number of coils wound on the package) during the time taken by the yarn guide to make a to and fro traverse. This to and fro traverses of the yarn guide from left to right and back from right to left is known as double traverse.

$$\text{Traverse ratio} = 2 \times \text{Wind}$$

2.5.3 Angle of Wind and Coil Angle

Angle of wind (θ) is the angle made by the yarn with the sides of the package (Figure 2.4). If surface and traverse speeds are V_s and V_t, respectively, then

$$\tan \theta = \frac{V_t}{V_s} \qquad (2.1)$$

Coil angle (α) is the angle made by the yarn with the axis of the package (Figure 2.4). The coil angle and angle of wind are complementary angles as they add up to 90°.

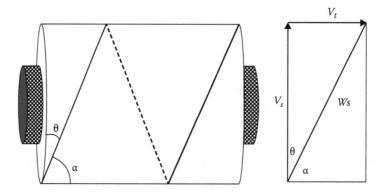

FIGURE 2.4   Angle of wind and coil angle.

2.5.3.1 Measuring the Angle of Wind

Figure 2.5 depicts a cheese which has parallel sides. The coils of yarn are wound over the surface of the cheese. The angle of wind can be measured in a very simple way. Keeping the cheese vertically, a straight line can be drawn on the package, parallel to the axis of the package. Now, if the yarns are unwound from the package, coloured dots will be visible on the yarn as shown by small circles in Figure 2.5. The distance between the two consecutive coloured dots indicates the length of one coil at this package diameter. The package diameter can be measured using a scale and then package circumference can be calculated. If the angle of wind is θ, then it can be calculated using the following expression.

$$\cos\theta = \frac{\text{Package circumference}}{\text{Length of a coil}}$$

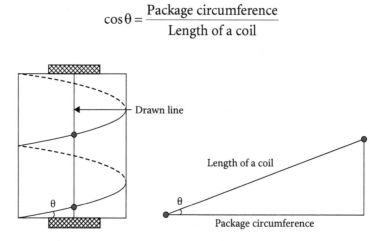

FIGURE 2.5   Measurement of angle of wind.

2.5.4 Winding Speed

It is the speed at which the yarn is being laid on the package. It is an indicator of productivity of winding machine. It has two components, namely surface speed of the package (V_s) and traverse speed (V_t). The net winding speed can be obtained by the resultant vector of surface speed and traverse speed (Figure 2.4).

$$\text{Winding speed} = W_s = \sqrt{\text{Surface speed}^2 + \text{Traverse speed}^2}$$

$$= \sqrt{V_s^2 + V_t^2} \tag{2.2}$$

2.5.5 Cone Taper Angle

The half of the angle at the cone top is known as cone taper angle (φ) as shown in Figure 2.6. If the empty cone diameters at the base and tip are D and d, respectively and the height of the cone is h, then the following expression can be written:

$$\tan \varphi = \frac{D-d}{2h} \tag{2.3}$$

The values of cone taper angles could be 3°30′, 4°40′, 5°50′ and so on, and it can go up to 11°.

2.5.6 Scroll

Scroll (S) is the number of revolutions that a grooved drum makes to execute one double traverse. Grooved drums with different scroll values are available. It is a parameter of grooved drum, and it influences the wind per double traverse and angle of wind of package.

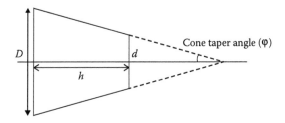

FIGURE 2.6   Cone taper angle.

2.6 WINDING MACHINES

Figure 2.7 depicts the simplified view of a winding machine. It has three main zones.

1. Unwinding zone

2. Yarn tensioning and clearing zone

3. Winding zone

In the unwinding zone, yarns are unwound from the supply package which is ringframe bobbin in most of the cases. Yarn balloon is formed due to the high-speed unwinding of yarn from the supply package. Unwinding tension varies continuously as the unwinding point shifts from the tip to the base and vice versa of a cop build ringframe bobbin. Besides, the height of the balloon also increases as the supply package becomes progressively empty.

In the second zone, tension is applied on the yarn using tensioners so that yarn is wound on the package with proper compactness. The objectionable yarn faults as well as other contaminations (coloured and foreign fibres) are also removed using optical or capacitance-based yarn clearer.

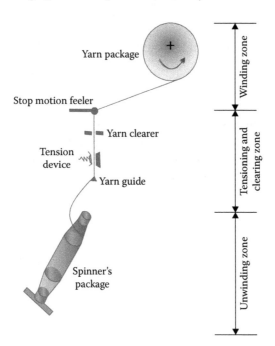

FIGURE 2.7   Zones of winding machine.

In the third and final zone, yarns are wound on the package by means of rotational motion of the package and traverse motion of the yarn guide. Based on the operating systems employed in the winding zone, two major winding principles have evolved.

2.7 CLASSIFICATION OF WINDING PRINCIPLES

Primarily there are three types of winding principles as given as follows:

1. Drum-driven or random winding

2. Spindle-driven or precision winding

3. Step-precision or Digicone winding

In a drum-driven wider, the package is driven by a cylinder by surface or frictional contact as shown in Figure 2.8. Traverse of yarn is given either by a grooved drum as shown at the bottom of Figure 2.8 or by a reciprocating

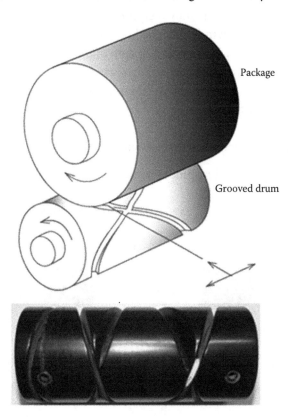

FIGURE 2.8   Drum-driven winder (top) and grooved drum (bottom).

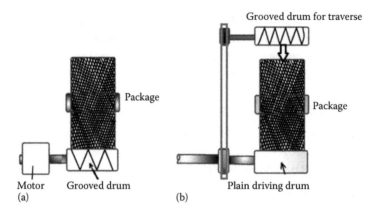

FIGURE 2.9 Types of drum-driven winder: (a) grooved drum and (b) plain drum.

guide. In the case of grooved drum, the drum performs the dual functions of rotating the package by surface contact and performing the traverse (Figure 2.9a). However, when plain drum is used, it just rotates the package and traverse is performed by reciprocating guide (Figure 2.9b).

In a spindle-driven winder, the package is mounted on a spindle which is driven positively by a gear system. If rpm of the spindle is constant, then the surface speed of the package will increase with the increase in package diameter. Therefore, principle-wise there could be two types of spindle-driven winder (Figure 2.10).

1. Constant rpm spindle–driven winder

2. Variable rpm–driven spindle winder

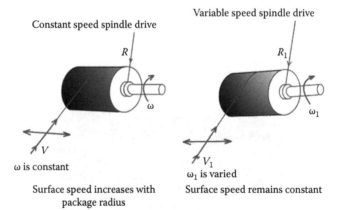

FIGURE 2.10 Spindle-driven winders.

In the case of the latter, the spindle rpm is reduced with the increase in package diameter in such a manner that the winding speed remains constant (Lord and Mohamed, 1982).

Spindle-driven winders are also known as precision winders as a precise ratio is maintained between the rpm of spindle and the rpm of traversing mechanism. This leads to maintaining a precise distance between the adjacent coils. The precision winders thus permit precise laying of coils on package and hence its name. Precision winders ensure a constant value of traverse ratio during package building. Precision winders are preferred for winding delicate yarns as the package is not rotated by the surface contact, and therefore, the possibility of yarn damage due to abrasion is lower as compared to that of drum-driven winders.

2.7.1 Drum-Driven Winders

Let us consider that the diameters of driving drum and package are D and d, respectively (Figure 2.11). The rpm of drum and package are N and n, respectively. D is constant, whereas d increases with time due to the building of package (formation of layers of coils). If there is no slippage between the drum and the package, then their surface speed will always be equal. So $N \times D = n \times d$. The drum rpm N is constant as the drum gets drive from the gear system. Thus n reduces as d increases with time.

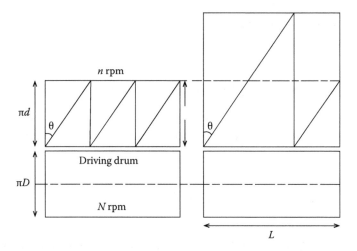

FIGURE 2.11   Principle of drum-driven winder.

In the case of grooved drum, as the drum rpm N is also constant, traverse speed and surface speed are also constant. Therefore, it produces constant angle of wind and winding speed.

Let L be the effective length of the drum and package.

Distance covered in one double traverse = $2L$

Number of drum revolutions required for one double traverse, that is scroll = S

N revolutions of drum takes 1 min

S revolutions of drum will take $\frac{S}{N}$ min

S revolution of drum is equivalent to one double traverse

So, time for one double traverse = $\frac{S}{N}$ min

$$\text{Traverse speed} = \frac{\text{Distance covered in one double traverse}}{\text{Time for one double traverse}} = \frac{2L}{S/N} = \frac{2LN}{S}$$

(2.4)

$$\tan\theta = \frac{V_t}{V_s} = \frac{2LN/S}{\pi DN} = \frac{2L}{\pi DS}$$

$$= \text{constant (as } L, D \text{ and } S \text{ are constant for a given drum)} \quad (2.5)$$

So, in a drum-driven winder, the angle of wind remains constant with the increase in the package diameter.

Now, as the package rpm is n, same (n) number of coils (wind) will be laid on the package every minute because one revolution of package creates one coil or wind on the package.

So

$$\text{Traverse ratio} = \text{Wind per double traverse}$$

$$= \frac{\text{Wind/min}}{\text{Double traverse/min}} = \frac{n}{\left(\dfrac{N}{S}\right)} = S \times \frac{n}{N}$$

$$= S \times \frac{D}{d} \quad (\text{as } N \times D = n \times d) \quad (2.6)$$

The package diameter d will increases with time during winding. So traverse ratio reduces with the increase in the package diameter as shown

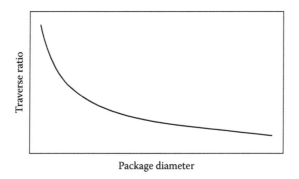

FIGURE 2.12   Package diameter *vs* traverse ratio.

in Figure 2.12. This leads to a 'patterning' problem in the case of a drum-driven winder.

$$\text{Winding speed} = \sqrt{\text{Surface speed}^2 + \text{Traverse speed}^2}$$

$$= \sqrt{(\pi DN)^2 + \left(\frac{2LN}{S}\right)^2} \tag{2.7}$$

$$= \sqrt{(\pi dn)^2 + \left(\frac{2LN}{S}\right)^2} \text{(considering no slippage)} \tag{2.8}$$

It is evident from the aforementioned expression that the winding speed remains constant during package building in the case of a drum-driven winder.

2.7.1.1 Cone Winding

Figure 2.13 depicts the situations when winding takes place on an empty and semi-full cone. The diameter of the empty cone at the base and tip is 6 and 4 cm, respectively. The diameter of the cylindrical grooved drum is 10 cm. When a cone and a cylindrical drum are in surface contact with each other and the latter is driving the former, there will be only one diameter on the cone where the surface speed of the two elements will be equal. Generally this point, which can change its position with package building, is located at a distance of one-third (1/3) of the cone height from the base (Koranne, 2013). Therefore, in this case the real contact between the cone and the drum is established at a point where the cone diameter is 5.33 mm.

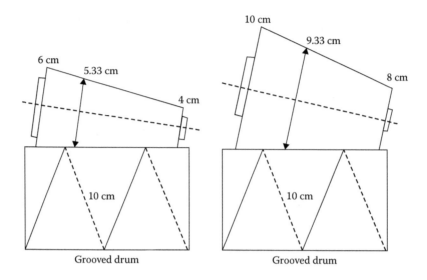

FIGURE 2.13 Winding on empty and semi-full cone.

TABLE 2.1 Change in Parameters during Cone Winding and Tip and Base

Parameter	Empty Cone	Semi-Full Cone
Number of cone revolution per drum revolution	$\dfrac{10}{5.33} = 1.88$	$\dfrac{10}{9.33} = 1.07$
Surface movement of the drum per revolution	$\pi \times 10 = 31.4$ cm	$\pi \times 10 = 31.4$ cm
Surface movement at the cone tip per drum revolution	$\pi \times 4 \times 1.88 = 23.61$ cm	$\pi \times 8 \times 1.07 = 26.87$ cm
Surface movement at the base per drum revolution	$\pi \times 6 \times 1.88 = 35.42$ cm	$\pi \times 10 \times 1.07 = 33.60$ cm
Ratio of surface speed between cone tip and drum	0.75	0.86
Ratio of surface speed between cone base and drum	1.13	1.07

The cone sections where the diameter is lower than 5.33 cm will have lower surface speed than that of the drum. On the other hand, the cone sections where the diameter is higher than 5.33 cm will have higher surface speed than that of the drum. Table 2.1 shows that the ratio of surface speeds of cone and drum at the tip and base is 0.75 and 1.13, respectively. This implies that the cone tip has 25% lower surface speed than that of the drum, whereas cone base has 13% higher surface speed than that of the drum.

As the cone diameter increases uniformly, the increment of diameter at the base and tip is same. This is shown at the right-hand side of Figure 2.13. The cone diameter has increased by 4 cm at the base as well as at the tip. Therefore, the present cone diameter at the base and tip is 10 and 8 cm, respectively. Assuming that the point of contact has remained unchanged, it is now located at cone diameter of 9.33 cm. Table 2.1 shows that for semi-full cone, the ratio of surface speeds of the cone and the drum at the tip and base is 0.86 and 1.07, respectively. This implies that the cone tip has 14% lower surface speed, whereas cone base has 7% higher surface speed than that of the drum. Therefore, the difference in surface speed between the drum and the cone is always higher at the tip than that of at the base. The difference in surface speeds between the drum and the cone causes abrasion between the yarn and the drum. This is more serious at the tip of the cone.

2.7.1.2 Patterning

If the traverse ratio (wind per double traverse) is an integer, then the yarn comes back to the same position on the package surface after one double traverse. This happens because the diameter of the yarn is minuscule and thus one double traverse does not cause significant increase in the package diameter. Therefore, in the next double traverse, the yarn coils are laid over the yarn coils which were laid in the previous double traverse. As a result, a ribbon develops on the package. This problem is known as 'patterning'. In the case of drum-driven winders, the traverse ratio reduces with the increase in package diameter. When the value becomes integer, the package becomes highly susceptible to patterning.

2.7.1.3 Path of Yarn on Cheese

Figure 2.14 shows the path of yarn, for one traverse, on a cheese having diameter d and length L. It is seen that the yarn has covered half of the package periphery when it has traversed half of the package length ($L/2$). This has been

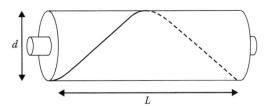

FIGURE 2.14   Path of yarn on a cheese.

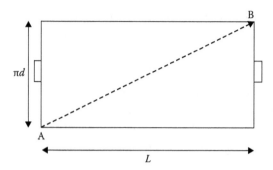

FIGURE 2.15   Path of yarn on a cheese after cutting.

depicted with a solid line in the figure. The yarn has made one complete coil (wind) when it has traversed the full length of the package. The path of the yarn in the second half of traverse has been shown with a broken line as it cannot be seen from the front view of the package. The situation can be visualised better if the package is cut along its axis and opened as shown in Figure 2.15.

In Figure 2.15, the yarn has started from A and moved to B. So it has made one coil and completed one traverse. If the process is continued, then the yarn will make two coils in one double traverse (i.e. traverse ratio = 2). The path of yarn in one double traverse is shown in Figure 2.16.

So the path of yarn in one double traverse can be simply written as ABB′A′. It must be kept in mind that on the package, A and A′ are the same points and similarly B and B′ are also the same points. It is understandable that after one double traverse the yarn has returned to the same position on the package from where it started its journey. So during the next double traverse, the yarn will be wound following the same path on the package. This will lead to the formation of ribbon on the package, which is known as patterning problem.

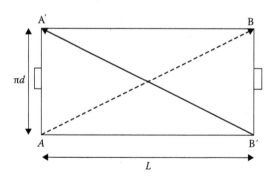

FIGURE 2.16   Path of yarn for traverse ratio 2.

2.7.1.4 Steps to Draw the Path of Yarn on Cheese

For any value of traverse ratio (wind per double traverse), the path of yarn on the cheese can be drawn by using the following steps.

- Traverse ratio (wind per double traverse) = x.

- So wind per traverse = $x/2$ and traverse per wind = $2/x$.

- Divide the opened package in to two equal parts (as the numerator is 2) in the vertical direction and x number of equal parts (as the denominator is x) in the horizontal directions. This will create some smaller rectangles within the opened package.

- Draw the diagonals for the small rectangles, starting from bottom-left corner.

- When one coil is complete, shift the winding point from the upper parallel line to the lower parallel line and vice versa.

- Reverse the direction of traverse when one traverse is complete.

If traverse ratio is 3 then the value of wind per traverse will be 3/2. So the value of traverse per wind will be 2/3. As per the aforementioned steps, the opened package has to be divided into two equal parts in vertical direction and three equal parts in horizontal direction. This is shown in Figure 2.17. The yarn will move from A to B to complete one coil. B and B′ are the same point on the package. The yarn will then move from B′ to C to complete one traverse. It can be seen from Figure 2.17 that while making one traverse, the yarn completes one and a half coils. Then the direction of traverse will change and now the yarn will move from C to B (right to left). Finally the yarn will complete the

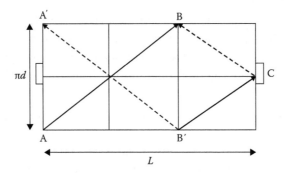

FIGURE 2.17 Path of yarn for traverse ratio 1.5.

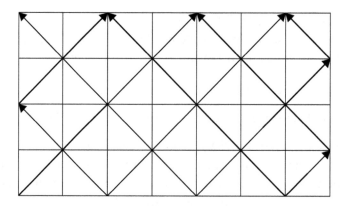

FIGURE 2.18   Path of yarn for traverse ratio 3.5.

double traverse by moving from B′ to A′. Here, the yarn comes back to the starting point (A) after only one double traverse.

Figure 2.18 depicts the path of yarn for a traverse ratio of 7/2 or 3.5. It should be noted that yarn comes back to the starting point after two complete double traverses, that is after laying seven coils. By extending this analogy, it can be inferred that if the traverse ratio is 15/4 or 3.75, then the yarn will come back to the starting point after four double traverses, that is after laying 15 coils. However, as the number of double traverse increases, the number of coils formed on the package also increases, which leads to the increase in package diameter. So the yarn actually comes back to a different point precluding the possibility of patterning. Therefore, patterning is prevalent when the traverse ratio is integer or having values like 1.5 or 2.5.

2.7.2 Spindle-Driven Winders

The schematic diagram of a spindle-driven winder is shown in Figure 2.19. The spindle carrying the package is rotating at n rpm. A and B (representing the respective tooth numbers) are the two gears responsible for transmitting the rotational motion from the spindle to traversing mechanism.

If these gears (A and B) are not changed then the ratio of spindle speed (rpm) and traverse speed (number of traverse/ min) remains the same and therefore the value of traverse ratio remains constant. However, the angle of wind changes during the package building, which can be understood from the following expressions.

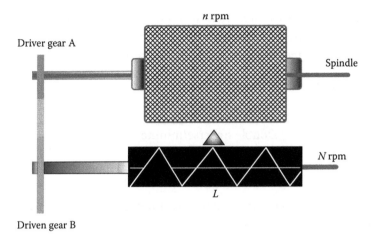

FIGURE 2.19 Drive of spindle-driven winder.

Let R be the number of double traverse made by the traversing device per minute and L be the length of the package.
So

$$\tan\theta = \frac{\text{Traverse speed}}{\text{Surface speed}}$$

$$= \frac{V_t}{V_s} = \frac{2LR}{\pi dn}$$

Now,

$$\text{rpm of the traversing drum} = N' = n \times \frac{A}{B}$$

If the traverse is given by a grooved drum which requires S revolutions for one double traverse then,

$$\text{Double traverse/minute} = R = \frac{N'}{S} = n \times \frac{A}{B} \times \frac{1}{S} \tag{2.9}$$

So

$$\tan\theta = \frac{2LR}{\pi dn} = \frac{2L}{\pi d} \times \frac{n \times \dfrac{A}{B} \times \dfrac{1}{S}}{n} = \frac{2L}{\pi d} \times \frac{A}{B} \times \frac{1}{S} \propto \frac{1}{d} \tag{2.10}$$

As the package diameter d *increases with the package building, the angle of wind (θ) decreases. It is also understood from the aforementioned expression that d tan θ remains constant for spindle-driven winders.*

Traverse ratio = Wind/double traverse

$$= \frac{\text{Wind/minute}}{\text{Double traverse/minute}} = \frac{n}{R} = \frac{n}{n \times \dfrac{A}{B} \times \dfrac{1}{S}} = \frac{B \times S}{A}$$

= Constant

As S is fixed for a given drum, the traverse ratio remains constant during the package building in a spindle-driven winder.

Figure 2.20 depicts the two situations with low and high package diameters. The traverse ratio is same in both the cases. However, the angle of wind has reduced from θ to α.

$$\text{Winding speed} = \sqrt{(\pi dn)^2 + (2LR)^2} \qquad (2.11)$$

The winding speed will increase with the increase in package diameter *d* if spindle speed *n* is constant. To keep the winding speed constant, the spindle speed must be reduced with the increase in the package diameter. It must be kept in mind that when spindle speed is reduced, the traverse speed (2*LR*) also reduces proportionately as the traversing mechanism is driven from the spindle.

In the case of a precision winder, the angle of wind reduces as the package diameter increases. This may lead to a problem as indicated in Figure 2.21. The theoretical traverse length is determined by the distance between the two extreme positions (left and right) of the yarn guide. Therefore, it remains unchanged during the building of yarn package. However, as the angle of wind is higher at the beginning, the yarn takes very sharp turn

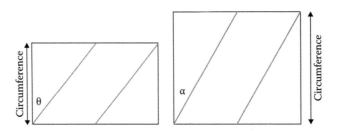

FIGURE 2.20   Change in angle of wind in precision winder.

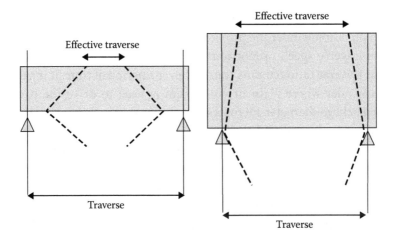

FIGURE 2.21   Change in effective traverse in precision winding.

at the extreme points and thus the effective traverse becomes less. As the package diameter increases, the angle of wind reduces and thus the yarn takes less sharp turns at the extreme points and thus effective traverse increases. Besides, the yarn has a tendency to slip at the reversal points and this propensity is more when the angle of wind is more, that is the package diameter is less. The increase in effective traverse with the increase in the package diameter may create tapered sides of precision wound package. This is shown in Figure 2.22.

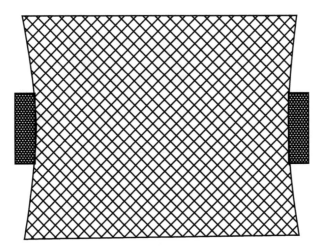

FIGURE 2.22   Sides of a precision wound package.

2.7.3 Step Precision Winder or Digicone Winder

In step precision winder the problem of patterning is prevented by changing the traverse speed proportionately with the package speed (rpm) so that the traverse ratio remains constant over a period of time. It is a drum-driven winder where plain driving drum is used to drive the package. As the package diameter increases, the package rpm decreases. The traverse speed is also reduced at the same rate through a cone drum system. However, after a certain time the traverse speed is raised back to the original value in one step, thereby moving quickly from one convenient value of traverse ratio to another. The schematic representation of the drive system of step precision winder is shown in Figure 2.23.

The plain driving drum gets motion directly from the motor. However, the motion goes to the traverse guide through a pair of cone drums. As the package rpm reduces, the belt connecting the two cone drums are shifted towards the left side in a controlled manner. Therefore, the actuating

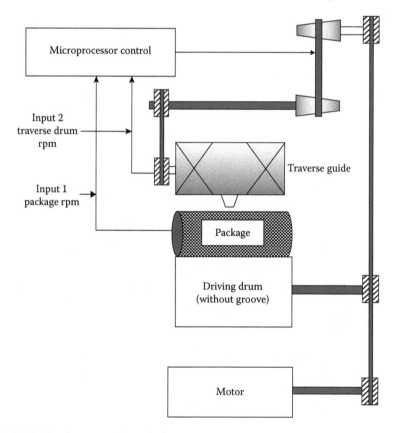

FIGURE 2.23   Step precision winder.

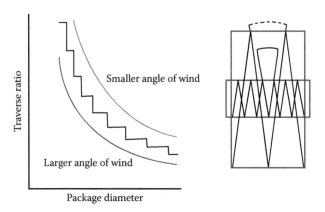

FIGURE 2.24   Change in traverse ratio with package diameter in Digicone winder.

diameter of the driver cone drum reduces, whereas it increases for the driven cone drum. So the speed of traversing system decreases and the traverse ratio remains constant over a period of time. However, the traverse speed cannot be reduced continuously as it will reduce the winding speed and angle of wind continuously. Therefore, after a certain time the connecting belt of the cone drums is shifted towards the right to restore the original value of traverse speed. Therefore, the traverse ratio, governed by the following equation, reduces in steps from one convenient value to the other.

$$\text{Traverse ratio} = \frac{\text{rpm of package}}{\text{Double traverse/min}}$$

Figure 2.24 depicts the change in traverse ratio with the increase in the package diameter. The angle of wind (θ) also changes by a small amount (1°–2°) when traverse speed is gradually reduced. However, it regains the original value when the traverse speed is raised back to the original value.

Table 2.2 summarizes the differences between three winding principles.

2.8   GAIN

Gain is the distance by which the winding point has to be shifted after one double traverse for avoiding patterning. Linear gain is measured in the direction perpendicular to package axis as shown in Figure 2.25. Traverse ratio basically quantifies the number of package revolution within a certain time (one double traverse). Therefore, linear gain cannot be added or subtracted with the traverse ratio. However, linear gain can be divided by the package circumference to obtain revolution gain, which can be added or subtracted with traverse ratio (Booth, 1977).

TABLE 2.2   Comparison of Winding Principles

Parameter	Drum-Driven	Spindle-Driven	Digicone
Angle of wind	Remains constant	Decreases with the increase in package diameter	Varies within a very small range
Traverse ratio	Decreases with the increase in package diameter	Remains constant	Remains constant for some time and then decreases in step
Winding speed	Remains constant	Generally increases with package diameter	Reduces slowly due to the reduction in traverse speed and then increases to the original value
Package density	Increases drastically at the zone of ribbons	Increases with the increase in package diameter	Density does not change with package diameter

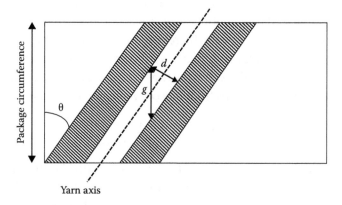

FIGURE 2.25   Gain for a cheese.

If winding takes place in such a manner that the coils wound in two consecutive double traverse are physically touching each other, then the gain can be expressed by the following equation.

$$\text{Linear gain} = g = \frac{\text{Yarn diameter}}{\sin \theta} \tag{2.12}$$

where θ is angle of wind.

$$\text{Revolution gain} = \frac{\text{Linear gain}}{\text{Package circumference}} \tag{2.13}$$

2.9 PIRN WINDING

Pirns are the yarn packages used within the shuttle to supply the yarn for pick insertion during weaving. The dimension of the shuttle is restricted by the shed geometry and the strain imposed on the warp yarns during shedding operation. The dimension of the pirn is governed by the dimension of the shuttle. Thus the pirn has to be a long and thin package. In contrast to cone winding, where the supply packages (ringframe bobbins) are smaller and the delivery packages (cones or cheeses) are bigger, the supply packages (generally comes) are bigger than the delivery package (pirn) in pirn winding. As the yarns have already been cleaned from slubs and other objectionable faults, no yarn clearing is required during pirn winding.

The winding principle of pirn is different than that of cones and cheeses. If a cross-wound package is made then there will be lot of tension variations during weaving. On the other hand, the parallel would package will give the problem of instability. Therefore, pirns are made by overlapping short, conical and cross-wound sections as shown in Figure 2.26.

The base of the empty pirn is generally conical. The pirn winding starts from the conical base and progressively proceeds towards the tip of the pirn. The distance travelled in one stroke of traverse is known as chase length. One layer of coils is laid on the conical base during the forward as well as during the return movement of the traversing mechanism.

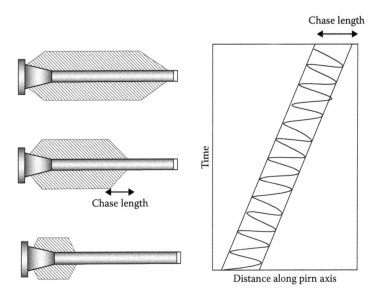

FIGURE 2.26   Stages of pirn winding.

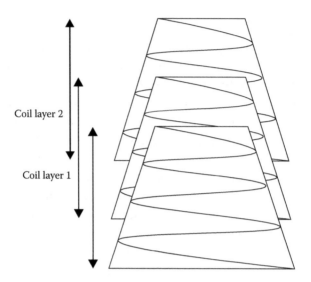

FIGURE 2.27   Building of a pirn.

Thus the conicity of the package is maintained and the tip of the cone, formed by the coils of yarn, slowly proceeds towards the tip of the pirn. The process can be visualised as if one plastic cup (having cone shape) is placed over another and the process is continued to build a tall cylindrical column. This is depicted in Figure 2.27. For the ease of visibility, large gaps have been maintained between two cones of coils.

Pirns may generally be described in the following categories or types.

- Plain tapered pirn

- Pirn with partly formed (half) base

- Pirn with full base

In the case of plain tapered pirn, the pirn body has continuous tapering as shown in Figure 2.28. In half base and full base pirns, the main body is cylindrical. However, the base is conical with higher cone angle in the case of the latter.

If the full and empty pirn diameter is D and d respectively, L is the chase length and α is the chase angle, then the following expression can be written.

$$\tan \alpha = \frac{D-d}{2L} \tag{2.14}$$

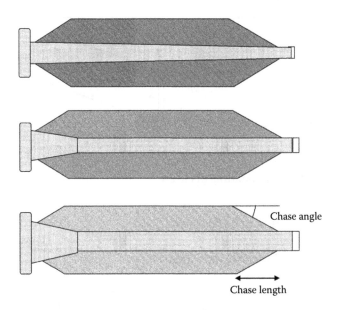

FIGURE 2.28   Plain, half base and full base pirns.

The chase angle depends on the type of yarn. For spun yarns, the chase angle could be as high as 15°–18°. However, for filament yarns with low friction, it could be as low as 6°–10°.

2.10 CONDITIONS FOR UNIFORM PACKAGE BUILDING

2.10.1 Cheese Winding

Uniform building of package is imperative in winding process. The package should have very uniform density. In the following part, the conditions for uniform building of cheese have been derived (Banerjee and Alagirusamy, 1999) based on the following assumption.

Assumption: Length of yarn wound per unit surface area of the package should be constant for uniform building of package.

The path of yarn on a cylindrical cheese is shown in the left-hand side of Figure 2.29. The schematic representation of the yarn path after cutting the package is shown in the right-hand side of Figure 2.29.

Current diameter of the package is d, length of the package is L and θ is angle of wind.

$$\text{Length of one coil of yarn} = AC = \frac{\pi d}{\cos \theta}$$

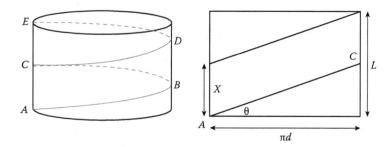

FIGURE 2.29 Uniform building of cheese.

$$\text{Number of coils in one traverse} = \frac{L}{X} = \frac{L}{\pi d \times \tan\theta}$$

Length of yarn/ surface area

$$= \frac{\text{Total length of yarn wound at diameter } d}{\text{Total surface area of package at diameter } d}$$

$$= \frac{\text{Length of one coil} \times \text{Number of coils at diameter } d}{\text{Total surface area of package at diameter } d}$$

$$= \frac{\dfrac{\pi d}{\cos\theta} \times \dfrac{L}{\pi d \times \tan\theta}}{\pi d L}$$

$$= \frac{1}{\pi d \sin\theta} \tag{2.15}$$

So $d \sin\theta$ must be kept constant for uniform building of the cheese.
    In the case of drum-driven winder,

$$\tan\theta = \frac{\text{Traverse speed}}{\text{Surface speed}} = \frac{V_t}{\pi n d}$$

where d and n are package diameter and rpm, respectively.
    So

$$V_t \cos\theta = \pi n d \sin\theta \tag{2.16}$$

For uniform package building $d \sin\theta$ should be kept constant.
    For drum-driven winder $n \times d = \text{constant}$
    So

$$n \propto \frac{1}{d}$$

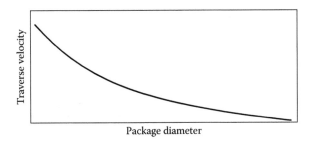

FIGURE 2.30   Package diameter *vs* traverse velocity.

So

$$V_t \cos\theta \text{ should be} \propto \frac{1}{d}$$

For drum-driven winders with grooved drums, θ remains constant during package building. Therefore, V_t should be reduced with the increase in the package diameter d to attain uniform building as depicted in Figure 2.30. However, for drum-driven winder, θ is constant provided the ratio of V_t and V_s are constant. But if V_t is reduced (keeping V_s constant, which is possible if traversing mechanism is separate from grooved drum), then θ will also reduce and $\cos\theta$ will increase. So V_t has to be reduced in such a manner that the product of V_t and $\cos\theta$ changes proportionately with $1/d$.

For spindle-driven winders, package rpm, that is n is constant.

So $V_t \cos\theta$ should be kept constant for uniform package building.

In spindle-driven winders, θ reduces (even when V_t is constant) as the package diameter (d) increases. So V_t has to be reduced accordingly as reducing V_t will have further bearing on θ.

2.10.2 Conditions for Uniform Cone Winding

In the case of cone, the diameter of package reduces as the yarn traverses from the base to the tip as shown in Figure 2.31. Surface speed of the cone is less in the tip part as compared to that of base part. Therefore, the situation becomes more complicated than that of cheese winding. It is important to maintain the conditions so that the diameters at the base and at the tip increase at the same rate. The condition for uniform increase of cone diameter has been derived in the following part.

Two sections of the cone having diameters d_1 and d_2 are being considered. Let w_1, v_1 and s_1 be the winding, traverse and surface speeds,

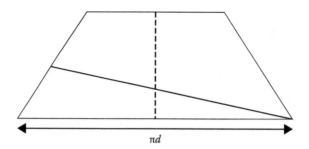

FIGURE 2.31   Yarn path on a cone.

respectively, at cone section diameter d_1. Similarly, let w_2, v_2 and s_2 be the winding, traverse and surface speeds, respectively, at cone section diameter d_2. For the analysis, a small time interval Δt has been considered.

Length of yarn wound per unit surface area at cone section diameter d_1

$$= \frac{w_1 \Delta t}{\pi d_1 v_1 \Delta t}$$

Length of yarn wound per unit surface area at cone section diameter d_2

$$= \frac{w_2 \Delta t}{\pi d_2 v_2 \Delta t}$$

For uniform increase in cone diameter, the boundary condition is

$$\frac{w_1 \Delta t}{\pi d_1 v_1 \Delta t} = \frac{w_2 \Delta t}{\pi d_2 v_2 \Delta t} \quad \text{or} \quad \frac{w_1}{w_2} = \frac{d_1 v_1}{d_2 v_2}$$

It is known that

$$\tan \theta = \frac{\text{Traverse speed}}{\text{Surface speed}} = \frac{v}{s}$$

Therefore,

$$\tan \theta_1 = \frac{v_1}{s_1} \quad \text{and} \quad \tan \theta_2 = \frac{v_2}{s_2}$$

and

$$w_1^2 = s_1^2 + v_1^2 \quad \text{and} \quad w_2^2 = s_2^2 + v_2^2$$

So

$$\frac{w_1^2}{w_2^2} = \frac{s_1^2 + v_1^2}{s_2^2 + v_2^2} = \frac{v_1^2\left(1+\dfrac{s_1^2}{v_1^2}\right)}{v_2^2\left(1+\dfrac{s_2^2}{v_2^2}\right)} = \frac{v_1^2\left(1+\cot^2\theta_1\right)}{v_2^2\left(1+\cot^2\theta_2\right)}$$

$$= \frac{v_1^2\sin^2\theta_2}{v_2^2\sin^2\theta_1}$$

From boundary condition

$$\frac{w_1}{w_2} = \frac{d_1 v_1}{d_2 v_2}$$

So

$$\left(\frac{w_1}{w_2}\right)^2 = \left(\frac{d_1 v_1}{d_2 v_2}\right)^2 = \frac{v_1^{\,2}\sin^2\theta_2}{v_2^{\,2}\sin^2\theta_1}$$

or

$$d_1^2\sin^2\theta_1 = d_2^2\sin^2\theta_2 \quad \text{or} \quad d\sin\theta = \text{constant}$$

It has been shown earlier that $V_t\cos\theta = \pi dn\sin\theta$.

Here, only one traverse is being considered and therefore cone rpm (n) will remain constant. For uniform increase of cone diameter, $d\sin\theta$ should also remain constant. Therefore, $V_t\cos\theta$ should be constant during one traverse from base to the tip of the cone. As the yarn moves towards the tip, the d reduces, so θ increases. As θ increases, $\cos\theta$ reduces. So it is needed to increase V_t in such a way that the $V_t\cos\theta$ remains constant because increase of V_t will have influence on θ also.

2.10.2.1 Example of Uniform Cone Winding

A cone is having varying section diameter from base to tip. However, at any instance, the entire cone revolves at the same rpm (n). Even if the traverse speed (V_t) is constant, the winding speed changes due to the change in surface speed as the winding point moves from base to the tip. This has been demonstrated pictorially in Figure 2.32. The surface speed of package

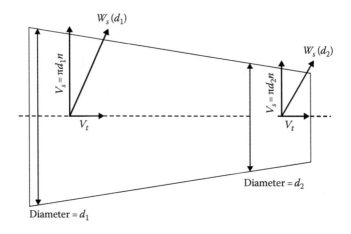

FIGURE 2.32   Winding at two different cone section diameters.

reduces as the winding point shifts towards the tip of the cone as $d_1 > d_2$. Therefore, the winding speed near the base, that is $Ws (d_1)$, is greater than that of near the tip that is $Ws (d_2)$. Let us consider two strips having unit width (1 cm) at cone section diameters d_1 (10 cm) and d_2 (8 cm) as shown in Figure 2.33. The length of yarn wound within these strips is a and b, respectively (shown by solid portions of inclined lines). So for uniform increase of cone diameter, the following condition must be maintained.

$$\frac{a}{b} = \frac{d_1}{d_2}$$

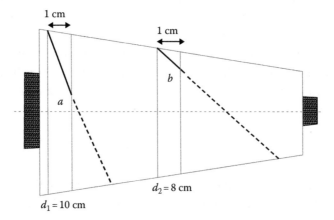

FIGURE 2.33   Uniform building of cone.

Now, if the length of yarn wound within the strip of cone section diameter d_1 is 3 cm, then the length of yarn wound within the strip at cone section diameter d_2 can be calculated as follows:

$$b = a \times \frac{d_2}{d_1} = 3 \times \frac{8}{10} = 2.4 \text{ cm}$$

As it has been considered that the width of each strip is 1 cm, the angle of wind at different cone section diameter can be calculated as follows:

$$\sin \theta_A = \frac{1}{3}; \quad \text{So } \theta_A = 19°27'$$

$$V_t \text{ (at diameter } d_1\text{)} = V_s \tan \theta_A = \pi \times 10n \times \tan 19°27' = 11.09n$$

$$\sin \theta_B = \frac{1}{2.4}; \quad \text{So } \theta_B = 24°37'$$

$$V_t \text{ (at diameter } d_2\text{)} = \pi \times 8n \times \tan 24°37' = 11.52n$$

Now, considering that the maximum and minimum cone section diameters are 12 cm and 6 cm, respectively, the angle of wind, traverse speed, surface speed and winding speed at different cone section diameters can be calculated as shown in Table 2.3. It is seen that the angle of wind increases towards the tip of the cone. The traverse speed increases whereas the surface speed reduces towards the tip of the cone. However, the magnitude of reduction in surface speed is much higher than the magnitude of increase in traverse speed towards the tip of the cone. Therefore, the overall winding speed reduces towards the tip of the cone. It should be noted that the value of $d\sin\theta$ is almost constant irrespective of the cone section diameter (Booth, 1977).

TABLE 2.3   Calculation of Angle of Wind and Speeds at Various Cone Diameters

Cone Section Diameter (cm)	Angle of Wind	Traverse Speed (cm/min)	Surface Speed (cm/min)	Winding Speed (cm/min)	$d \sin \theta$
12	16°8'	10.90n	$\pi \times 12 \times n = 37.7n$	39.2	3.36
10	19°27'	11.09n	$\pi \times 10 \times n = 31.4n$	33.3	3.34
8	24°37'	11.52n	$\pi \times 8 \times n = 25.1n$	27.6	3.35
6	33°45'	12.60n	$\pi \times 6 \times n = 18.8n$	22.6	3.34

2.10.3 Grooves on Winding Drums

Generally grooves on the driving drum of a winder have constant angle and pitch. However, grooves can have varying angle and pitch in the case of drums used for cone winding. The grooves can have increasing pitch from the base to the tip of the cone as shown in Figure 2.34. In this case, the first revolution of the drum causes one-third traverse (L/3) across the length of the package. The second revolution of the drum causes two-third traverse (2L/3) across the length of the package. This implies that as the yarn moves from the base to the tip of the cone, the traverse speed increases as shown in Figure 2.35. These drums are known as accelerated grooved drums.

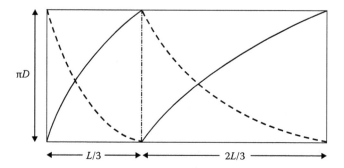

FIGURE 2.34   Grooves on accelerated grooved drum.

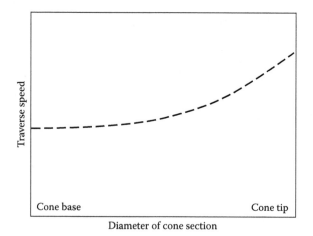

FIGURE 2.35   Traverse speed *vs* cone section diameter.

2.11 YARN TENSIONING

The primary objective of yarn tensioning is to build yarn packages with adequate compactness. Higher yarn tension than the optimum will result in a tighter package and vice versa. If there is any portion of yarn which has very low tensile strength (untwisted part of yarn), then it will not be able to sustain the applied winding tension and as a result the yarn will break. This will lead to momentary stoppage in winding operation. However, this will preclude the possibility of yarn breakage in the subsequent processes like warping. As a rule of thumb, yarn tension in winding is around 1 cN/tex.

2.11.1 Types of Tensioning Devices

Two types of tensioning devises are used in winding machines.

1. Additive type or disc-type tensioner

2. Multiplicative-type tensioner

In the case of the former, the yarn is passed through two smooth discs, one of which is weighted with the aid of small circular metallic pieces (Figure 2.36). The weights can be changed easily so that the tension in the output yarn can be adjusted as per the requirement. If T_1 and T_2 are the tension (cN) in the input and output yarns, respectively, and w is the weight (cN) applied on the top disc and μ is the coefficient of friction between the yarn and metal disc, then the following expression can be written:

$$T_2 = T_1 + 2\mu w \qquad (2.17)$$

The factor 2 appears in this expression as the lower disc also offers reaction forces on the yarn as shown in Figure 2.36.

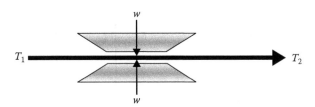

FIGURE 2.36 Additive-type tensioner.

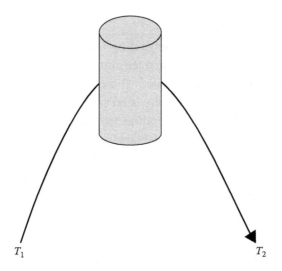

FIGURE 2.37   Multiplicative-type tensioner.

In the case of multiplicative type tensioner, the yarn is passed around a curved or cylindrical element as shown in Figure 2.37.

Here, the relationship between input (T_1) and output (T_2) tensions can be expressed as follows:

$$T_2 = T_1 \times e^{\mu\theta} \tag{2.18}$$

where θ is the angle of warp (in radian) of the yarn around the tensioning element.

Figure 2.38 depicts two situations with angle of warps of $\pi/2$ and π.

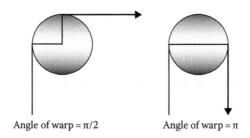

Angle of warp = $\pi/2$ Angle of warp = π

FIGURE 2.38   Different angle of warps.

2.11.2 Relation between Input and Output Tensions in Multiplicative Tensioner

Let it be assumed that the yarn is passing over a curved surface, which is considered to be the part of a circle (Figure 2.39). The contact region between the curved surface and the yarn has created a small angle $d\theta$ at the centre of the assumed circle. The yarn tensions at the input and output sides are T and $T + dT$, respectively. Let it be assumed that the difference between the horizontal components of T and $T + dT$ will balance the frictional resistance, which will depend on the coefficient of friction between the yarn and tensioner (μ) and the resultant (P) of vertical components of T and $T + dT$.

Balancing the forces, the following equations can be written:

Resolving and balancing the vertical components,

$$P = (T + dT)\sin\frac{d\theta}{2} + T\sin\frac{d\theta}{2}$$

$$= 2T\sin\frac{d\theta}{2}$$

$$\left(\begin{array}{l} \text{the product of } \sin\dfrac{d\theta}{2} \text{ and } dT \text{ can be} \\[2mm] \text{ignored as both of them are small} \end{array}\right)$$

$$= 2T \times \frac{d\theta}{2} \left(\text{as } \frac{d\theta}{2} \text{ is small, } \sin\frac{d\theta}{2} = \frac{d\theta}{2} \right)$$

$$= T \, d\theta$$

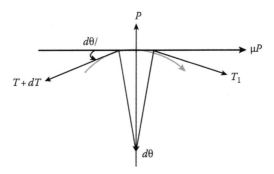

FIGURE 2.39 Schematic representation of yarn passing over a curved surface.

Resolving and balancing the horizontal components,

$$\mu P = (T + dT)\cos\frac{d\theta}{2} - T\cos\frac{d\theta}{2}$$

$$= dT\cos\frac{d\theta}{2} = dT \left(\text{as } \frac{d\theta}{2} \text{ is very small, } \cos\frac{d\theta}{2} \approx 1 \right)$$

Now,

$$\mu P = dT$$

or

$$\mu T\, d\theta = dT$$

or

$$\mu \int_0^\theta d\theta = \int_{T_1}^{T_2} \frac{dT}{T}$$

or

$$\mu\theta = \log\frac{T_2}{T_1}$$

or

$$\frac{T_2}{T_1} = e^{\mu\theta} \tag{2.19}$$

So, if the input and output tensions are T_1 and T_2, respectively, and angle of contact between the yarn and the tensioner is θ, then equation 2.18 describes their relationship.

2.11.3 Tension Variation during Unwinding from Cop Build Packages

During unwinding of yarns from cop build packages (e.g. ringframe bobbin, pirn), short-term and long-term tension variation is noticed. Short-term tension variation arises due to the movement of yarn unwinding point from the tip to the base and vice versa (Figure 2.40a). On the other hand, long-term tension variation occurs due to the change in height of the balloon formed between the unwinding point and the yarn guide (Figure 2.40b).

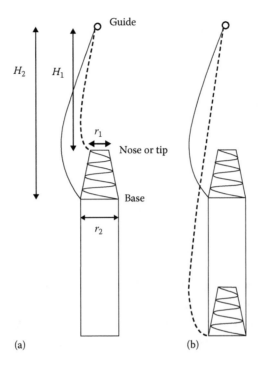

FIGURE 2.40 Tension variations during unwinding (a) short term (b) long term.

The empirical equation for unwinding tension is given as follows:

$$\text{Unwinding tension} \propto mv^2 \left[C_1 + C_2 \left(\frac{H}{r} \right)^2 \right] \qquad (2.20)$$

where
 H is balloon height
 r is package radius (varies between tip and base)
 m is mass per unit length of yarn
 v is unwinding speed

In the case of short-term tension variation, one layer of coil is unwound and the yarn withdrawal point moves from the tip to the base. As a result, both the H and r increase. However, the proportionate change in r is higher as compared to that of H. From Figure 2.40a, it can be seen that $H_1/r_1 \gg H_2/r_2$. Therefore, in every cycle, when the yarn withdrawal point moves

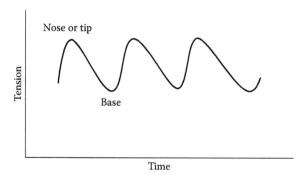

FIGURE 2.41   Short-term tension variation.

from the tip to the base, the unwinding tension reduces. This is shown in Figure 2.41.

However, over a long period of time, successive conical layers of yarns are removed from the package and thus the conical section of yarns moves towards the base of the pirn. Therefore, the balloon height increases, resulting in progressive increase in mean unwinding tension. This is shown in Figure 2.42.

In modem winding machines, the unwinding tension is maintained at constant level by continuously measuring and controlling it. Saurer-Schlafhorst offers such a system called Autotense in winding machine Autoconer X5. The working principle of Autotense is shown in Figure 2.43. Yarn is wound on the delivery package at constant speed. The yarn tension sensor measures the tension acting on the yarn continuously and provides feedback to the control system. The control system either increases or decreases the tensioner pressure based on the level of yarn tension. For example, as the bobbin starts to exhaust, the balloon height increases,

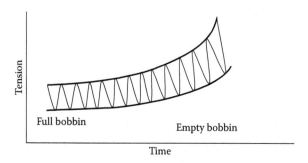

FIGURE 2.42   Short-term and long-term tension variation.

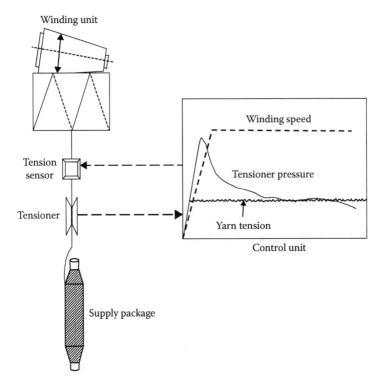

FIGURE 2.43   Automatic control of winding tension.

causing increase in yarn tension. Therefore, the tensioner pressure is reduced to keep the final winding tension at constant level. This is shown inside the box within Figure 2.43.

2.12 YARN CLEARING

2.12.1 Objectives of Yarn Clearing

The objective of yarn clearing is to remove objectionable faults from the yarn. Ideally all the faults present in the yarn should be removed during the yarn clearing operation. However, a compromise is needed and only those faults which have potential to disrupt the subsequent operations or spoil the fabric appearance are attempted for removal during the winding operation. The compromise is done due to the following reasons:

- Removal of yarn faults during winding is associated with the machine stoppages, which reduce the winding machine efficiency.

- When a yarn fault is removed, the two ends of yarn are joined again by knotting or splicing, which actually introduces a new blemish in the yarn as the strength and appearance of the knotted or spliced region are not at par with those of the normal region of yarn.

2.12.2 Principles of Measurement

Two principles are used in modern winders for the identification of yarn faults.

- Capacitance principle

- Optical principle

Both the principles have their inherent advantages and limitations. Capacitance type systems are based on the measurement of yarn mass at a given test length. In contrast, the optical systems are based on the diameter measurement.

Figure 2.44 depicts the principle of a capacitance-based yarn clearing system. The yarn is passed at a constant velocity through two parallel plate capacitors. The expression of capacitance of a parallel plate capacitor is as follows:

$$\text{Capacitance} = C = \frac{\varepsilon A}{d} = \frac{k\varepsilon_0 A}{d} \tag{2.21}$$

where
A is the area of the capacitor plates
d is the distance between the plates
ε is the permittivity of the medium present between the capacitor plates
ε_0 is the permittivity of vacuum (8.85×10^{-12} Farad/m)
k is the dielectric constant of the medium

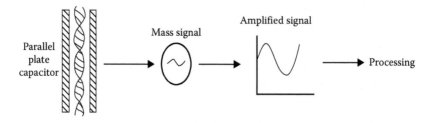

FIGURE 2.44 Principle of capacitance-based measurement.

When the yarn passes through the parallel plates, the equation will take the following form.

$$C = \frac{A}{\left(\dfrac{d_1}{\varepsilon_1} + \dfrac{d_2}{\varepsilon_2}\right)} = \frac{A\varepsilon_0}{\left(\dfrac{d_1}{k_1} + \dfrac{d_2}{k_2}\right)} \qquad (2.22)$$

where

d_1 is the thickness of material 1 (yarn)

d_2 is the thickness of material 2 (air)

ε_1 is the permittivity of material 1

$\varepsilon_2 (\approx \varepsilon_0)$ is the permittivity of material 2

k_1 is the dielectric constant of material 1

k_2 is the dielectric constant of material 2

Based on the mass of yarn present within the parallel plate capacitor, the capacitance changes, which is converted into mass unevenness signal to identify the faults. The dielectric constant of water is 80, whereas for textile fibres it is around 2–5 and for air it is nearly 1. Thus the measurement is highly sensitive to the presence of moisture and therefore conditioning of samples in standard atmospheric conditions is of paramount importance.

Figure 2.45 represents the principle of optical-based yarn clearing system. The emitter emits light, and the receiver detects it and converts to proportional electrical signal. The light received by the receiver will obviously depend on the diameter of the yarn passing between emitter and receiver.

It should be considered that 10% deviation in diameter will actually cause 21% deviation in the mass because for circular yarn cross-section, mass is proportionate with square of diameter. Therefore, principally capacitance-based measurements are more sensitive to deviation than the optical-based measurements. However, some of the faults which may not be detected by capacitance-type testers can be detected by optical-type testers. The fault may be a low twisted region or a hole within the yarn structure (Figure 2.46). If the mass per unit length is same, then the

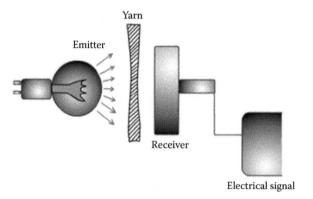

FIGURE 2.45 Principle of optical-based measurement.

FIGURE 2.46 Yarn with a hole within the structure.

capacitance-type tester will not detect this yarn fault, although there is a deviation in diameter which can produce fabric defects. This can be detected by optical-type testers as higher yarn diameter will reduce the amount of light received by the receiver. Optical-type testers are perceived to be better in terms of predicting fabric appearance.

2.12.3 Yarn Imperfections and Yarn Faults

Yarn blemishes are broadly categorized under two heads.

1. Frequently occurring type or yarn imperfections

2. Seldom occurring type or yarn faults

Yarn imperfections appear very frequently in the spun yarns. Generally, they do not pose serious threat to the subsequent processes or fabric appearance. Therefore, yarn imperfections are never attempted for removal in yarn winding. Frequently occurring faults are measured by yarn unevenness testers and expressed by the frequency of occurrences per km. The three types of yarn imperfections are given as follows:

1. Thick places (mass exceeds by at least +50% of the nominal mass)

2. Thin places (mass is lower by at least −50% of the nominal mass)

3. Neps (mass exceeds by +200% of the nominal mass with reference length of 1 mm)

The cumulative number of thick place, thin place and neps per km of yarn is known as imperfections or IPI value.

Yarn faults are seldom occurring defects. They can adversely affect the efficiency of warping and weaving due to frequent yarn breakages. In addition, they can severely damage the appearance of the yarn fabric. Yarn faults are evaluated by Classimat testers, which have different variants like Classimat 3, Classimat 4 and Classimat 5. The number of fault types is 23, 33 and 45 in Classimat 3, Classimat 4 or Classimat 5, respectively, depending on the length and mass of the fault. Yarn faults are generally expressed by the number of occurrences per 100 km. Figure 2.47 shows the fault matrix of Classimat 3 where mass and length of faults are indicated in the vertical and horizontal directions, respectively.

The Classimat 3 faults are classified under three major categories:

1. Short thick faults:	A1–D4
2. Long thick faults:	E, F and G
3. Long thin faults:	H1, I1, H2, I2

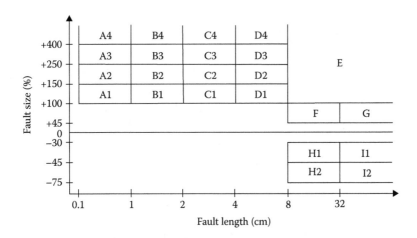

FIGURE 2.47   Matrix of Classimat 3 faults.

The classifications A, B, C and D correspond to fault reference lengths of 0.1–1, 1–2, 2–4 and 4–8 cm, respectively. The fault size % indicates the percentage increase in the fault mass varying from +100% to more than +400%, corresponding to diameter increase of 41% and 123%. This results in 16 fault types, known as short thick faults, with A1 being the shortest in length and the smallest in diameter and D4 being the longest in length and the largest in diameter. Spinners' double refer to a long thick fault (E) whose length exceeds 8 cm and mass exceeds +100%. F and G are also long thick faults as their mass exceeds the nominal level by +45% and length is between 8 and 32 cm and greater than 32 cm, respectively. Within the long thin category, H faults (H1 and H2) have length between 8 and 32 cm whereas I faults (I1 and I2) are longer than 32 cm. Among the short thick faults, A4, B4, C4, D4, C3 and D3 are generally considered as objectionable faults as their length and deviation from the nominal mass are very high. Now, A3, B3, C2 and D2 are also considered as objectionable faults for high-quality yarns.

Research work by Aggarwal et al. (1987) showed that C3, C4 and all D Classimat faults, even after sizing, have lower tenacity, reduced extensibility and poor abrasion resistance. These Classimat faults introduce a very high frequency of low strength and low elongation portions into the yam, which in turn causes warp breaks.

Figure 2.48 shows the fault matrix of Classimat 4. In addition to the 23 faults of Classimat 3, there are 10 additional faults which are classified under two new categories, which are as follows:

1. Very short thick fault (A0, B0, C0 and D0)

2. Short thin faults (TB1, TC1, TD1, TB2, TC2, TD2)

For very short thick faults (A0–D0), the mass exceeds the nominal mass by +70%. The reference length of these faults is same with A–D, which has been mentioned earlier. For TB1, TC1 and TD1, the mass is lower than the nominal mass by 30%–45% whereas for TB2, TC2 and TD2 this is 45%–75%.

Figure 2.49 shows the fault matrix of Classimat 5, where the number of faults in the matrix has gone up to 45 (23 standard classes and 22 additional classes with respect to Classimat 3).

2.12.3.1 Causes of Classimat Faults

The short thick faults (A1–D4) are caused either by the deficiency in raw material and preparatory processes (blowroom and carding) or due to the drafting problems. If a diagonal line is drawn joining A4 and D1, then the

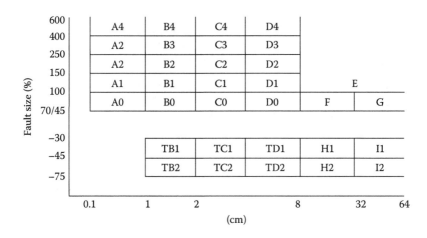

FIGURE 2.48 Matrix of Classimat 4 faults.

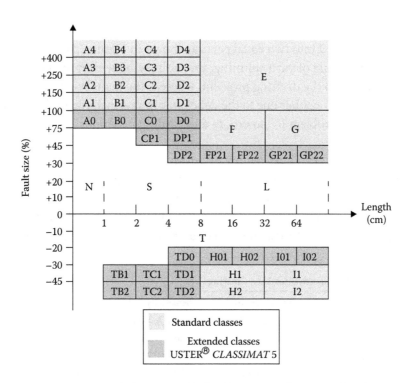

FIGURE 2.49 Matrix of Classimat 5 faults. (Courtesy of Uster Technologies AG, Uster, Switzerland.)

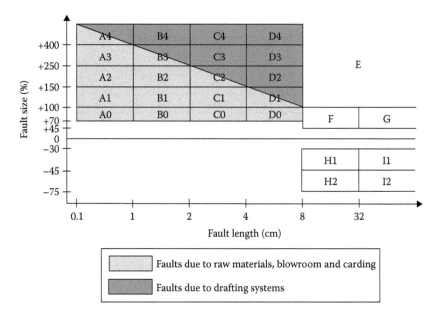

FIGURE 2.50 Causes of short thick faults.

matrix is divided into two equal parts as shown in Figure 2.50. According to the thumb rule of yarn spinning, the faults lying within the upper triangle are due to the drafting problem, whereas the faults lying within the lower triangle are either due to the deficiency in raw material or due to the fibre opening problem in blowroom and carding machines.

2.12.3.2 Settings of Yarn Clearing Channels

In modern winding machines, multiple channels are used for the clearing of yarn faults. The details of these channels are presented in Table 2.4.

In winding machine, the user has the flexibility to adjust the settings of these channels to optimise the yarn clearing in terms of fault removal and

TABLE 2.4 Channels for Yarn Clearing

Type of Channel	Target Removal	Sensitivity (%)	Reference Length (cm)
N	Very short thick faults	+100 to +500	0.1–1
S	Short thick faults	+50 to +300	1–10
L	Long thick faults	+10 to +200	1–200
T	Long thin faults	−10 to −80	10–200
C	Count deviations	±5 to ±80	1280

Source: Courtesy of Uster Technologies AG, Uster, Switzerland.

machine stoppages (winding cuts/km). The setting depends on the yarn count as well as quality (carded or combed).

Typical setting in the industry for S, L and T channels can be as follows:

- S channel: Mass +140% to + 200% and length 1.5–2 cm

- L channel: Mass +40% to + 50% and length 40–50 cm

- T channel: Mass −30% to −40% and length 40–50 cm

2.12.4 Removal of Foreign and Coloured Fibres

The contamination in cotton fibre is one of the most serious causes of quality problems. The contamination includes jute, polypropylene, husk, leaf, hair and paper. Presence of foreign fibres such as polypropylene or polyethylene in cotton yarns creates quality problem due to uneven dyeing. Modern blowrooms are equipped with systems like Vision Shield®, which ejects the contaminated portions of cotton flocks using sophisticated image processing technology. Some of the spinning mills prefer manual cotton sorting in addition to the automated systems. The electronic yarn clearers in modern winding machines remove not only the objectionable faults but also the foreign and coloured fibres by the foreign fibre channel. Some of the systems used in winding machines to remove foreign fibres are given as follows:

- Loepfe Zenit Yarn Master®

- Uster Quantum 2

Loepfe Zenit uses triboelectric effect, which is a type of contact electrification, for foreign fibre identification. According to this principle, certain materials become electrically charged after they come into contact with a different material. Materials have a specific order of the polarity of charge separation when they are touched or abraded with another object. A material which acquires higher position in the triboelectric series will attain positive charge when touched by a material positioned near the bottom of the series. The further away the two materials are from each other on the series, the greater the charge transferred.

Uster Quantum presents the distribution of foreign fibres in a matrix based on % deviation in appearance and fault length (Kretzschmar and Furter, 2008). In the case of fine setting, the faults in the classes B1, B2, C1, D1 and E1 are further divided into subclasses as shown in Figures 2.51 and 2.52. No counting

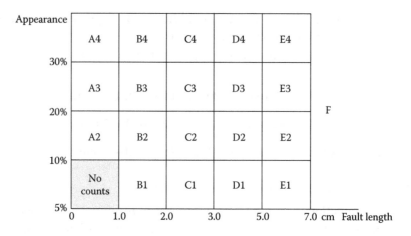

FIGURE 2.51 Coarse setting in Uster Quantum 2. (Courtesy of Uster Technologies AG, Uster, Switzerland.)

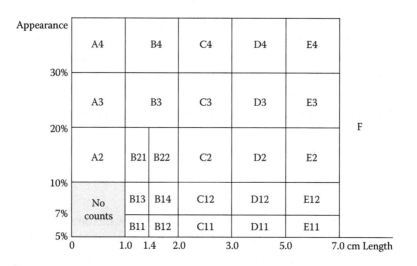

FIGURE 2.52 Fine setting in Uster Quantum 2. (Courtesy of Uster Technologies AG, Uster, Switzerland.)

is done in the lowest block of the left-hand side as the number of faults in this category is very high. During winding, only the serious types of foreign fibre faults are attempted for removal. Otherwise the winding efficiency may drop drastically. Experiments with foreign fibre removal systems have revealed that their clearing efficiency is dependent on the colour of foreign fibre and shade depth. The clearing efficiency is found to be higher for red followed by blue and yellow. The clearing efficiency was found to increase with shade depth.

2.13 SPLICING

Splicing is the process by which two ends of the yarns are joined. In most of the winding machines dealing with spun yarns, pneumatic splicers are used. Thermo-splicers can be used for thermoplastic yarns. Robotic arms aided with air suction bring the two ends of the yarns inside the splicing chamber. Then compressed air is jetted to create turbulence inside the chamber so that the yarn ends are untwisted. Then some fibres are removed from the yarn ends to create wedge shape. Jetting of compressed air is done again to twist the two superimposed ends of yarns (Figure 2.53). Splicing parameters such as air pressure, overlapping length of two yarn ends and duration of air blowing influence the quality of splice. Coarser yarns and yarn spun from fibres with higher torsional rigidity necessitate more air pressure as untwisting and twisting become difficult. Splicing introduces a less severe fault in the yarn and the appearance of spliced portion of yarns is checked by making the yarn appearance board. The quality of spliced yarn is evaluated by using the parameters such as retained splice strength and splice breaking ratio as defined as follows:

$$\text{Retained splice strength (\%)} = \frac{\text{Breaking strength of spliced yarn}}{\text{Breaking strength of parent yarn}} \times 100$$

$$\text{Splice breaking ratio} = \frac{\text{No of breaks in spliced zone } (\pm 1 \text{ cm})}{\text{Total number of tests}}$$

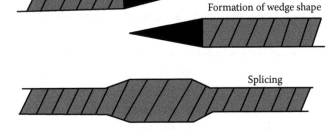

Two ends of yarn

Formation of wedge shape

Splicing

FIGURE 2.53   Steps of yarn splicing.

Higher retained splice strength (85%–90%) and lower splice breaking ratio imply good splicing performance.

The performance of a splicer is also evaluated by the parameters such as clearing efficiency and splice factor. As knotting device is not used in modern winders like Autoconer, the term 'knot factor' may be replaced with 'splice factor'. Higher clearing efficiency and lower splice factor (close to 1) signify good performance of a splicer.

Clearing efficiency

$$= \frac{\text{Number of objectionable faults removed}}{\text{Total number of objectionable faults originally present in yarn}} \times 100$$

$$\text{Splice factor} = \frac{\text{Number of yarn clearer related breaks}}{\text{Number of objectionable faults removed}}$$

2.14 SOME IMPORTANT ISSUES OF WINDING

2.14.1 Yarn Winding for Package Dyeing

Yarns are often dyed in package form. Yarn packages intended for dyeing should have certain characteristics to facilitate uniform dyeing within and between packages. The density of the package should be relatively low but uniform. For cotton yarn, the density of package should be around 0.35–0.40 g/c.c. for optimum dyeing performance. Low density will ensure better penetration and flow of dye liquor across the yarn layers. On the other hand, uniform density will ensure that no preferential channel of fluid flow is developed. Some of the requirements of packages are as follows:

- Density variation within and between the packages should be less than ±2.5%.

- Dye package outer diameter variation should be within ±1 mm.

Drum-driven or random winders are not preferred for packages intended for dyeing due to patterning problem which may increase the package density drastically. In the case of precision winder, the angle of wind reduces as the package diameter increases. Thus the package density increases towards the outer side of the package. In the case of step precision winder or Digicone winder, the angle of wind varies marginally during winding (±1°–2°). Thus the density of the package remains nearly constant. Digicone wider produces the best packages for dyeing.

In principle, the density of the package can be varied by the following ways:

- Changing the angle of wind

- Changing the distance between neighbouring yarn coils within a layer

- Changing the pressure between the package and the drum

- Changing the winding tension

High angle of wind causes lower package density (Figure 2.54). The higher the distance between the neighbouring yarns within the same layer, the lower the package density (Figure 2.55). In Figure 2.54, the path of yarn from left to right is shown with solid lines and that from right to left with broken lines. The angle of wind and distance between the neighbouring

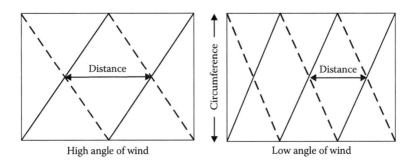

FIGURE 2.54   Angle of wind and package density.

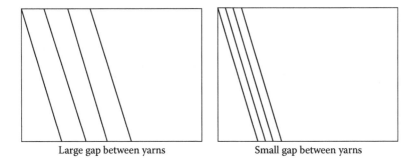

FIGURE 2.55   Gap between yarns and package density.

yarns cannot be changed independently in drum or spindle-driven winders. For example, in a drum-driven winder, if the angle of wind is changed using a winding drum having different scroll (S) value, the distance between the two neighbouring yarns within the same layer will also change. This can be understood from Figure 2.54. Besides, in a drum-driven winder, the distance between the neighbouring yarns, within a layer, increases with the increase in the package diameter, whereas it remains constant in the case of spindle-driven or precision winder (Figure 2.56).

Higher pressure between package and winding drum and higher winding tension increases the package density. In some of the modern winders, the pressure between the package and winding drum is reduced as the package diameter increases so that the outer layers do not become denser than the inner layers. Most of the modern winders have automatic tension control system, which minimizes the tension variation that arises due to the change in diameter or due to change in the unwinding point on the supply packages. For a given winding speed, this system ensures controlled increase or decrease of winding tension based on the input tension.

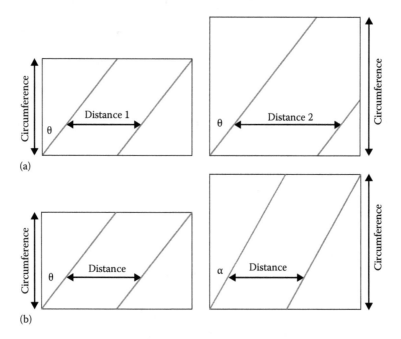

FIGURE 2.56 Distance between neighbouring yarns at different package diameters. (a) Change in distance between neighbouring yarns in drum-driven winder. (b) No change in distance between neighbouring yarns in spindle-driven winder.

2.14.2 Yarn Waxing

The yarns intended to be used for knitting requires reduced metal to yarn coefficient of friction. During the knitting process, the yarn makes contact with feeder eye, needle, sinkers and many other machine components. Higher coefficient of friction can increase the tension acting on the yarns during the knitting process. Besides, the protruding hairs may get entangled with the needle latch, causing improper opening and closing of the latter. Therefore, waxing is done during the winding. Generally the add-on of wax on the yarn is around 0.1%–0.3%. The choice of type of wax is very important as it influences the add-on %. Soft waxes cause more add-on than the hard waxes. Over-application of wax on yarn is also detrimental as it can increase the coefficient of friction.

2.14.3 Defects in Wound Packages

Some of the major defects observed in wound packages are as follows:

- Ribbon or pattern
- Stitches
- Soft tip or base
- Slough off

The ribbons are formed when the yarn coils of two successive layers rest over or are very closely placed to one another as shown in Figure 2.57. This generally happens in a drum-driven winder when the traverse ratio becomes an integer. Thick ridges are formed due to patterning and it mars the appearance of the package. Besides, the density of the package at the ribbon becomes very high, which causes problem in unwinding and dyeing. Patterning is prevented in some of the drum-driven machines either by momentarily lifting the package from the drum or by reducing the contact pressure between the package and drum and thereby creating some intended slippage. This causes change in yarn path from the anticipated one and thus patterning problem is avoided. In other machines, the anti-patterning device continuously varies the speed of the traverse gear. Thus, the winding angle is changed continuously, preventing the possibility of patterns to a large extent.

Stitches are formed when the yarn is wound beyond the boundary of the package. It may happen due to improper traverse guidance of the yarn at the edges of the package. If the winding tension is low then the yarn may exceed

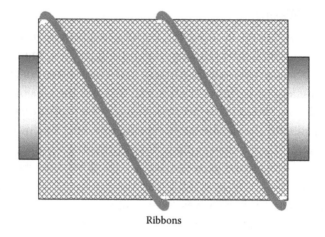

Ribbons

FIGURE 2.57   Package with ribbon or patterning defect.

the boundary line up to which it is supposed to travel. The improper contact between the package and the winding drum may also cause stitches.

If the pressure between the package and the drum along the line of contact is not uniform, then the package may have different density at the tip and base regions. Higher contact pressure will lead to higher package density and vice versa. The alignment of the winding drum and the package must be correct to ensure uniform density along the axis of the package. The position of the traverse guide should be exactly at the centre of the winding drum so that the length of the yarn between package and the traverse guide remains the same at the two extreme positions of traverse.

Slough off is the problem of removal of multiple coil from the package during high-speed unwinding. If the package density or gain is not adequate, then slough off may occur when the wound package will be used in warping section.

2.14.4 Winding and Yarn Hairiness
The yarn hairiness increases due to winding because of the following:

- Abrasion of yarn with various machine parts

- Transfer of one fibre from one section of yarn to the other

Research by Rust and Peykamian (1992) has shown that yarn hairiness increases by a greater amount if the winding speed is high. During winding, redistribution of twist takes place. The twist flows to the regions with lower yarn diameters. This causes fibre transfer and increase in yarn hairiness. Though the average helix angle remains unaltered after winding, the variation in twist angle reduces after winding, which supports the fact that twist rearrangement takes place during winding.

NUMERICAL PROBLEMS

2.1 The yarn tensioning system shown in Figure 2.58 is being used in a winding machine. The input and output tensions are 10 and 98 cN, respectively. Coefficient of friction between the yarn and metal is 0.3. If disc (additive)-type tensioners A and B are identical, then calculate the weights used in tensioners A and B.

Solution:

Let the weight of the discs be NcN.

Input tension $= T_0 = 10$ cN

Tension in yarn after tensioner A $= T_1 = T_0 + 2\mu N = 10 + 2\mu N$

Tension after the first multiplicative tensioner $= T_2 = T_1 e^{\mu\theta}$

$$\text{The angle of warp} = \theta = 90° = \frac{\pi}{2}$$

Tension after the second multiplicative tensioner $= T_3 = T_2 e^{\mu\theta'} = \left(T_1 e^{\mu\theta}\right) e^{\mu\theta'}$

$$= T_1 e^{2\mu\theta} \left(\text{as } \theta = \theta'\right)$$

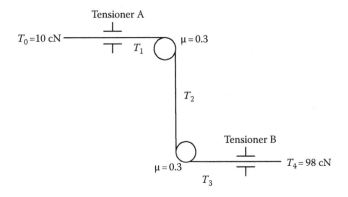

FIGURE 2.58 Yarn tensioning system.

$$\text{Tension after tensioner } B = T_4 = T_3 + 2\mu N$$

$$= T_1 e^{2\mu\theta} + 2\mu N$$

Now,

$$T_1 e^{2\mu\theta} + 2\mu N = 98$$

or

$$\left(10 + 2\mu N\right) e^{2\mu\theta} + 2\mu N = 98$$

or

$$(10 + 0.6N) e^{0.3\pi} + 0.6N = 98$$

or

$$2.56(10 + 0.6N) + 0.6N = 98$$

or

$$N = 33.9 \approx 34$$

So the weight of each of the discs is 34 cN.

2.2 The bare diameter of a spindle-driven cylindrical package is 5 cm. The spindle speed is 2000 rpm and traverse velocity is 100 m/min. Determine the following:

(a) Winding speed and angle of wind at the start
(b) Surface speed and angle of wind when the package diameter becomes double

Solution:

(a) Here

Spindle rpm (n) = 2000
Traverse speed (V_t) = 100 m/min

d = bare diameter of cylindrical package = 5 cm

Surface speed $(V_s) = \pi dn$

$$= \frac{\pi \times 5 \times 2000}{100} \text{ m/min}$$

$$= 314.29 \text{ m/min}$$

$$\tan \theta_1 = \frac{V_t}{V_s} = \frac{100}{314.29} = 0.318,$$

where θ_1 is the angle of wind
or
$\theta_1 = 17.65°$

$$\text{Winding speed} = \sqrt{\left(V_t^2 + V_s^2\right)}$$

$$= \sqrt{100^2 + 314.29^2}$$

$$= 329.81 \text{ m/min.}$$

So winding speed and angle of wind at the start are 329.81 m/min and 17.65°, respectively.

(b) When the package diameter becomes double, that is $d' = 10$ cm

Surface speed $(V_s') = \pi d'n$

$$= \frac{\pi \times 10 \times 2000}{100} \text{ m/min}$$

$$= 628.57 \text{ m/min}$$

Now,

$$\tan \theta_2 = \frac{V_t}{V_s'} = \frac{100}{628.57} = 0.159$$

or

$$\theta_2 = 9.04°$$

So surface speed and angle of wind when the package diameter is double are 628.57 m/min and 9.04°, respectively.

2.3 A cheese of 150 mm traverse length is wound on a drum-driven machine equipped with 75 mm diameter drum of 2.5 crossings. Calculate the winding speed and coil angle if the drum rpm is 3250.

Solution:

Figure 2.59 depicts the cheese and the winding drum.

$$\text{Surface speed of drum} \left(V_s\right) = \pi DN$$

where

  D is the diameter of drum = 75 mm = 7.5 cm
  N is the rpm of the drum = 3250

$$V_s = \frac{\pi \times 7.5 \times 3250}{100} = 766.07 \text{ m/min.}$$

$$\text{Traverse speed} \left(V_t\right) = \frac{2LN}{S}$$

where L is the traverse length = 150 mm = 15 cm

  Scroll (S), that is no. of drum revolution for double traverse = 2× crossings = 2 × 2.5 = 5

$$V_t = \frac{2 \times 15 \times 3250}{5 \times 100} = 195 \text{ m/min}$$

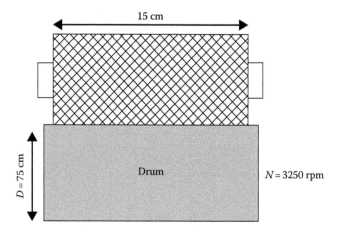

FIGURE 2.59 Cheese and winding drum.

So

$$\text{Winding speed} = \sqrt{(\pi DN)^2 + \left(\frac{2LN}{S}\right)^2}$$

$$= \sqrt{766.07^2 + 195^2}$$

$$= 790.5 \text{ m/min}$$

$$\tan\theta = \frac{V_t}{V_s} = \frac{195}{766.07} = 0.255$$

So
θ (angle of wind) = 14.3°
Therefore, the coil angle is 75.7°.

2.4 The diameter and length of the drum of a random winder are 100 mm and 300 mm, respectively. The winding speed is 1200 m/min. If the drum is rotating at 4000 rpm and crossing on drum is 3, then calculate the slippage %.

Solution:
The drum diameter (D) = 100 mm = 10 cm

Drum length (L) = 300 mm = 30 cm

Drum rpm (N) = 4000

No. of revolutions of drum per double traverse (S) = 3 × 2 = 6

Let the translation ratio between package surface speed and drum surface speed be x.
   Therefore,

$$\frac{\text{Package surface speed}}{\text{Drum surface speed}} = x$$

Slippage = (1 − x) and is applicable only for the surface speed part as there is no slippage in the traverse speed part. Considering the slippage, the expression of winding speed will be as follows:

$$\text{Winding speed} = \sqrt{(\pi DN \cdot x)^2 + \left(\frac{2LN}{S}\right)^2}$$

$$= \left[\left(\frac{\pi \times 10 \times 4000 \times x}{100}\right)^2 + \left(\frac{2 \times 30 \times 4000}{6 \times 100}\right)^2\right]^{\frac{1}{2}}$$

$$= 1200 \text{ m/min.}$$

or

$$1256.63x = 1131.37$$

or

$$x = 0.9$$

So slippage is 0.1 or 10%.

2.5 A cone with negligible cone taper angle and having maximum mean diameter of 200 mm is being built using a 50 mm diameter cylindrical winding drum rotating at 3000 rpm. The number of traverse per minute is 320. The cone was running at 960 rpm when the mean package diameter was 150 mm. What is the slippage % between drum and package? Assuming no change in the position of contact point between the drum and package during winding, calculate the cone diameters at which patterning may occur.

Solution:
Figure 2.60 depicts the drum and the cone.

The rpm of the cone (n) is 960 when the package diameter (d) is 150 mm = 15 cm.

Therefore,

$$\text{The slippage is} = 1 - \frac{\pi dn}{\pi DN}$$

where
  d is the mean package (cone) diameter
  D is the drum diameter
  N is the rpm of the drum

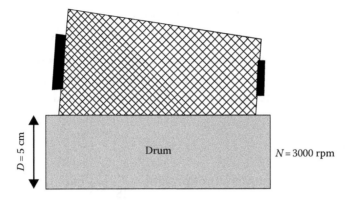

FIGURE 2.60　The drum and the cone.

$$\text{Slippage} = 1 - \frac{15 \times 960}{5 \times 3000}$$

$$= 0.4, \text{ that is } 4\%$$

Maximum cone diameter (mean) = 200 mm = 20 cm.

Patterning will occur when traverse ratio (n/N_t) will be integer, where N_t is the traverse/min.
Here

$$N_t = 320$$

So values of n will be = 320 × x where $x \subset K$ (x belongs to the set of natural numbers, i.e. K)

Now, considering that the slippage between the drum and cone to be constant,

$$n \times d = 960 \times 15 = 14400 = \text{constant}$$

So
320 × x × d = 14,400

Now, considering $x = 1$, the value of d becomes 45 cm, which is not acceptable because 45 cm > 20 cm (maximum mean package diameter).

The values of d at which patterning may occur are 15, 11.25, 9, 7.5 cm and so on. (corresponding values of x are 3, 4, 5 and 6).

2.6 What is the nearest value of traverse ratio to 3 to prevent patterning in a cheese when the package diameter is 5 cm and the angle of wind is 30°? The yarn is made up of cotton fibre and the yarn count is 20 Ne.

Solution:

$$\text{Diameter of cotton yarn} = \frac{2.54}{28\sqrt{\text{Ne}}}\ \text{cm}$$

$$= \frac{2.54}{28\sqrt{20}}\ \text{cm}$$

$$= 0.02\ \text{cm}$$

So

$$\text{Linear gain is } \frac{0.02}{\sin 30°} = 0.04\ \text{cm}$$

$$\text{Revolution gain} = \frac{\text{Linear gain}}{\pi d}$$

$$= \frac{0.04}{\pi \times 5}$$

$$= 0.0025$$

So the value of traverse ratio is 3 ± 0.0025.

2.7 (a) For a spindle-driven winder, the angle of wind is 20° when the package diameter is 10 cm. Determine the angle of wind when the package diameter is 15 cm. (b) In a drum-driven winder, the angle of wind is 30°. The drum having 5 cm diameter makes 5 revolutions for one double traverse. Calculate the length of the winding drum.

Solution:

(a) For spindle-driven winders,

$$\tan\theta = \frac{\text{Traverse speed}}{\text{Surface speed}}$$

$$= \frac{V_t}{V_s} = \frac{2LR}{\pi dn} = \frac{\text{constant}}{\pi dn}$$

where
  L is the traverse length, that is effective length of the drum
  R is the number of double traverse/min.
  d is the package diameter
  n is the package or spindle rpm

As, R/n and L are constants, so $d \tan\theta = $ constant.
  So

$$\tan\theta_2 = \frac{d_1 \tan\theta_1}{d_2} = \frac{10 \times \tan 20°}{15}$$

or

$$\theta_2 = 13.64°$$

So the angle of wind will be 13.64° at 15 cm package diameter.

(b) For drum-driven winders,

$$\tan\theta = \frac{Vt}{Vs} = \frac{2LN/S}{\pi DN} = \frac{2L}{\pi DS}$$

where
  L is the traverse length
  N is the rpm of drum
  S is the number of drum revolutions for one double traverse

So

$$\tan 30° = \frac{2L}{\pi \times 5 \times 5}$$

or

$$L = 22.68 \text{ cm.}$$

So the length of winding drum is 22.68 cm.

2.8 A precision winder with traverse length of 20 cm is operating at constant winding speed of 1000 m/min. The spindle rpm is 3000 when the package diameter is 10 cm. What will be the spindle rpm when the package diameter increases to 20 cm?

Solution:

$$\text{Constant winding speed} = w = 1000 \text{ m/min}$$

The spindle speed has to be reduced, with the increase in the package diameter, to maintain constant winding speed. The traverse speed ($2LR$) will also reduce as it is originating from spindle drive.

$$\text{Spindle rpm } (n_1) = 3000$$

$$\text{when the package diameter } (d_1) = 10 \text{ cm} = 0.1 \text{ m.}$$

$$\text{Traverse length} = L = 20 \text{ cm} = 0.2 \text{ m.}$$

$$\text{Increased package diameter } (d_2) = 20 \text{ cm} = 0.2 \text{ m.}$$

Let $(R/n) = k$ (constant, where R is the number of double traverse per minute)
or

$$R = nk$$

In case 1:

$$w^2 = (2LR)^2 + (\pi d_1 n_1)^2$$
$$= (2Ln_1 k)^2 + (\pi d_1 n_1)^2$$

So

$$w^2 = n_1^2 \left(4L^2 k^2 + \pi^2 d_1^2 \right)$$

or

$$1000^2 = 3000^2 (4 \times 0.04 \times k^2 + \pi^2 \times 0.01)$$

or

$$k^2 = \frac{0.0125}{0.16} = 0.0782$$

In case 2:

$$w^2 = n_2^2 \left(4L^2 k^2 + \pi^2 d_2^2 \right)$$

$$1000^2 = n_2^2 \left(4L^2 k^2 + \pi^2 d_2^2 \right)$$

$$= n_2^2 (4 \times 0.04 \times 0.0782 + \pi^2 \times 0.04)$$

So

$$n_2 = 1568 \text{ rpm.}$$

The spindle speed will be 1568 rpm when the package diameter is 20 cm.

2.9 A cheese of 20 cm traverse length is being wound on a precision winder with traverse ratio of 5/2. What should be the traverse ratio nearest to 5/2 for the prevention of patterning when the cheese diameter is 10 cm? The yarn is made of cotton fibre having packing fraction is 0.6 and yarn count is 4 Ne.

Solution:

The path of yarn is shown in Figure 2.61.

$$\text{Traverse ratio} = \frac{\text{Wind/minute}}{\text{Double traverse/minute}} = \frac{5}{2} = 2.5$$

Therefore, wind/traverse = 1.25. So 1.25 coils occupy 20 cm of traverse length. Therefore, one coil will occupy 16 cm of traverse length.

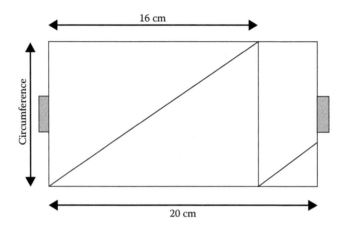

FIGURE 2.61 Path of yarn on cheese.

$$\tan\theta = \frac{\text{Traverse length occupied by one coil}}{\text{Package circumference}} = \frac{16}{\pi\times 10}$$

So

$$\theta = 27°$$

$$\text{Yarn diameter (cm)} = d = \frac{2.54}{28\sqrt{4}} = 0.045$$

$$\text{Linear gain } (g) = \frac{d}{\sin\theta} = \frac{0.045}{\sin 27°} = \frac{0.045}{0.454} = 0.099$$

$$\text{Revolution gain} = \frac{g}{\pi d} = \frac{0.099}{\pi\times 10} = 0.003$$

So, for avoiding patterning, traverse ratio should be 2.5 ± 0.003, that is 2.503 or 2.497.

2.10 A precision winder with constant spindle speed is being used for the production of cheese. The traverse length 150 mm and the traverse ratio is 6. If the minimum angle of wind is 8°, determine the maximum package diameter.

Solution:

Traverse length of cheese (L) = 150 mm = 15 cm

$$\text{Traverse ratio} = \frac{\text{Package rpm }(n)}{\text{Double traverse per min }(R)} = 6$$

In the case of precision winder, the angle of wind (θ) reduces as the package diameter (d) increases.

Therefore, θ_{min} corresponds to d_{max}

$$\tan\theta_{min} = \frac{V_t}{V_s} = \frac{2LR}{\pi d_{max} n}$$

$$= \frac{2\times15\times R}{\pi\times d_{max}\times n} \quad \text{(for spindle-driven or precision winders)}$$

or

$$\tan 8° = \frac{30}{\pi\times d_{max}}\times\frac{1}{6} = \frac{5}{\pi\times d_{max}}$$

or

$$d_{max} = \frac{5}{\pi\times\tan 8°} = 11.32 \text{ cm}$$

So maximum cheese diameter is 11.32 cm.

2.11 A cheese having 150 mm length, 50 mm bare diameter and 125 mm maximum diameter is being wound on two different machines as follows:

(a) A grooved drum rotary traverse winder where the drum diameter is 75 mm and the drum is making 5 revolutions per double traverse.

(b) A constant spindle speed precision winder where the spindle rotates 6 times per double traverse.

For machine (a) determine the traverse ratio at package diameters of 50 and 100 mm. For machine (b) find the angle of wind at diameters of 50 and 100 mm.

Solution:

Traverse length (L) = 150 mm =15 cm

Minimum diameter of the cheese (d_{min}) = 50 mm = 5 cm

Maximum diameter of the cheese (d_{max}) = 125 mm = 12.5 cm

(a) Drum diameter (D) = 75 mm = 7.5 cm

Revolution of drum per double traverse (S) = 5

$$\text{Traverse ratio} = S \times \frac{D}{d}$$

$$\text{When } d = 5 \text{ cm, traverse ratio} = 5 \times \frac{7.5}{5} = 7.5$$

$$\text{When } d = 10 \text{ cm, traverse ratio} = S \times \frac{D}{d} = 5 \times \frac{7.5}{10} = 3.75$$

So the values of traverse ratio are 7.5 and 3.75 at package diameters of 50 and 100 mm, respectively.

(b) $\tan \theta_1 = \dfrac{V_t}{V_s}$

or

$$\tan \theta_1 = \frac{2LR}{\pi dn},$$

here

$$\frac{n}{R} = 6$$

as the spindle rotates at 6 times per double traverse.

$$\tan \theta_1 = \frac{2 \times 15}{3.14 \times 5} \times \frac{1}{6} = 0.3184$$

or

$$\theta_1 = 17.65°$$

Again

$$\tan\theta_2 = \frac{V_t'}{V_s'} = \frac{2\times 15}{\pi\times 10}\times\frac{1}{6}$$

or

$$\theta_2 = 9.04°$$

So the values of angle of wind are 17.65° and 9.04° at package diameters of 50 and 100 mm, respectively.

REFERENCES

Aggarwal, A. K., Hari, P. K. and Subramanian, T. A. 1987. Evaluation of classimat faults for their performance in weaving. *Textile Research Journal*, 57: 735–740.

Banerjee, P. K. and Alagirusamy, R. 1999. *Yarn Winding*. New Delhi, India: NCUTE Publications, Indian Institute of Technology.

Booth, J. E. 1977. *Textile Mathematics*, Vol. III. Manchester, UK: The Textile Institute.

Koranne, M. 2013. *Fundamentals of Yarn Winding*. New Delhi, India: Woodhead Publishing Limited.

Kretzschmar, S. D. and Furter, R. 2008. Application report of Uster classimat quantum, Uster Technologies AG, Uster, Switzerland. http://www.uster.com. Accessed on 21st March 2016.

Lord, P. R. and Mohamed, M. H. 1982. *Weaving: Conversion of Yarn to Fabric*, 2nd edn. Merrow, UK: Merrow Technical Library.

Rust, J. P. and Peykamian, S. 1992. Yarn hairiness and the process of winding. *Textile Research Journal*, 62: 685–689.

Technical literature of Autoconer X5, Saurer-Schlafhorst, Germany. www.saurer.com. Accessed on 21st March 2016.

CHAPTER **3**

Warping

3.1 OBJECTIVES

The objective of warping is to convert yarn packages into a warper's beam having desired width and containing the requisite number of ends. Uniform tension is maintained on individual yarns during warping.

The yarns are wound on the warper's beam in the form of a sheet composed of parallel yarns each coming out from a package placed on the creel. A schematic view of the warping process is shown in Figure 3.1.

3.2 BEAM WARPING AND SECTIONAL WARPING

Two principal methods of warping are beam warping and sectional warping (Booth, 1977). Let us consider one hypothetical example to understand the various options of warping process (Banerjee, 2015).

A woven fabric roll of 2 m width and 10,000 m length is to be produced from warp yarn of 15 tex. There should be 40 yarns per cm in the fabric. Therefore, the total number of ends in the fabric will be 40 × 200 = 8000. Considering that 10 beams will be combined in sizing, the number of ends on each warper's beam will be 800. Ignoring the yarn crimp and wastage of yarns, the mass of a single end having 10,000 m length will be 150 g. Therefore, there will be the following options:

$P \times R$ indicates the number of packages used × number of warping runs

1. **800 × 1: Each cone should contain 10,000 m yarn, weighing 150 g.**

2. 400 × 2: Each cone should contain 20,000 m yarn, weighing 300 g.

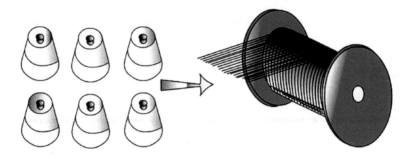

FIGURE 3.1   Simplified representation of warping process.

3. 200 × 4: Each cone should contain 40,000 m yarn, weighing 600 g.

4. 100 × 8: Each cone should contain 80,000 m yarn, weighing 1200 g.

5. **50 × 16: Each cone should contain 160,000 m yarn, weighing 2400 g.**

In the first option, 800 packages are required, whereas it is 400 in the second option and so on. Now, first and fifth options represent two extreme situations. Option 1 (direct warping or beam warping) can be executed when the lot being processed has significantly higher (15–20 times) length than that considered (10,000 m) in this example. Considering the mass of a full cone as 2.1 kg, if the ordered length is 140,000 m, then the entire cone (150 × 14 = 2100 g) will be consumed.

On the other hand, the fifth option (sectional warping or indirect warping) is practiced when multi-coloured warp patterns or specialty yarns are used for manufacturing customised products. In this case, the production planning officer does not see the possibility of repeat order in near future. Therefore, he or she wants to consume the entire package to minimise the wastage and inventory-carrying cost. Therefore, the beam is made section by section and the operation is repeated a large number of times to complete the entire width of the warper's beam. This is also followed by the beaming operation, when all the sections of warp are transferred from the drum to the flanged beam.

In synthetic filament weaving systems, each supply package contains a very high length of yarn. Such packages are therefore fairly expensive, and it is also very difficult to store such packages with unspent yarn. Hence, synthetic yarn weaving units prefer to opt for sectional warping.

For direct warping the typical length of lot can be as follows:

- 40 Ne: 1.6 lakhs m

- 50 Ne: 1.8 lakhs m

- 60 Ne: 1.9 lakhs m

3.3 COMPONENTS OF WARPING MACHINE

The major components of a warping machine are as follows:

- Creel (Figure 3.2)

- Control devices

- Headstock

Supply packages for warping are mounted on the creel. Creels are of three types: single end, magazine and travelling or swivelling. After the creel, the yarns pass through the guides and tensioners. Finally, the warp sheet is wound on the beam in the headstock section.

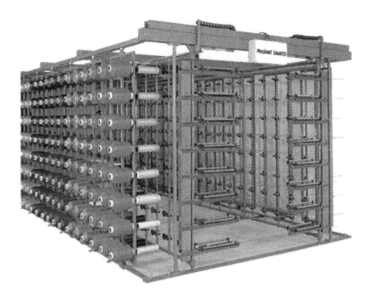

FIGURE 3.2   A simple creel of warping machine. (Courtesy of Prashant Group of Industries, Ahmedabad, India.)

3.3.1 Types of Creel

3.3.1.1 Single End Creel

In single end creel, one package position (package holder) of the creel is used for one end on the warper's beam. Single end creel can be of two types: truck creels and duplicated creels. The creel is movable in the case of the former, whereas the headstock is movable in the case of the latter (Lord and Mohamed, 1982). In the case of truck creel, when the packages from the running creel are exhausted, it is moved sideways and the reserve creel is moved into the vacant space (Figure 3.3). Thus, the time for removing a huge number of exhausted packages and replenishing them with full packages is saved. However, extra space is required for the reserve creel.

3.3.1.2 Magazine Creel

In magazine creel, the tail end of yarn from one supply package is tied with the tip of yarn of another neighbouring supply package. When the first package is exhausted, the yarn withdrawal from the second package begins automatically without machine stoppage. This is depicted in Figure 3.4. Thus the creeling time is reduced, which helps improve the running efficiency of warping machine. However, due to sudden change in unwinding position and tension variation associated with this, some of the yarns break during the transfer (known as transfer failure). Magazine creel has reduced capacity. If the creel has 1000 package holders, then the warp sheet can actually have 500 ends.

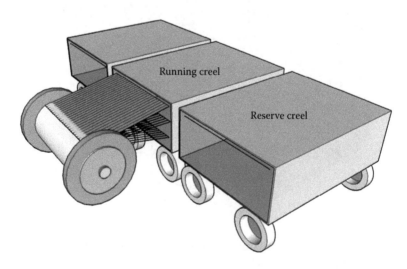

FIGURE 3.3   Single end truck creel.

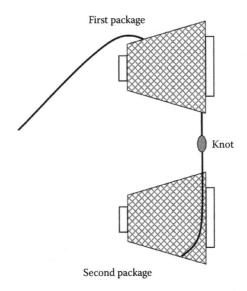

FIGURE 3.4 Magazine creel.

3.3.1.3 Travelling or Swivelling Creel

In swivelling creel, the pegs (package holders) with full packages can move from inside (reserve) position to the outside (working) position when the running packages are exhausted, thus saving considerable time. The operator replaces the exhausted packages with full packages when the machine is running. Figure 3.5 shows the swivelling creel (Lord and Mohamed, 1982).

3.3.2 Calculation for Warping Efficiency with Different Creels

Let us assume that warping operation is being carried out with the following particulars:

The yarn mass on full beam is 300 kg, number of ends is 500, yarn count is 30 tex, warping speed is 1000 m/min, cone mass is 2 kg, end breakage rate 0.1/100 end/1000 m, time to repair a break is 0.5 min, beam doffing time is 5 min, creeling time is 45 min/creel, headstock change time is 3 min/ beam and transfer failure is 2%.

The calculation which follows has been done considering the time required for various operations with respect to one warper's beam. Table 3.1 shows the time required for various operations.

The mass of yarn contributed by a package to one warper's beam
$$= 300/500 = 0.6 \, \text{kg}$$

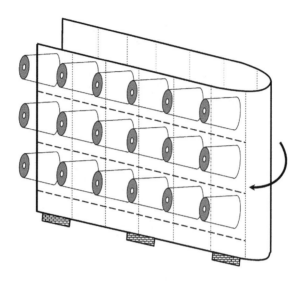

FIGURE 3.5 Swivelling creel.

TABLE 3.1 Time Required for Various Operations

Item	Calculation/Beam	Single End (min)	Duplicated (min)	Magazine (min)
Running time	20,000 m warp sheet	20	20	20
Repair time	0.5 min/break	5	5	5
Beam doffing	5 min/beam	5	5	5
Creeling time	45 min/creel	15	0	0
Headstock change	3 min/headstock	0	3	0
Transfer failure	2%			1.67
Total time		45	33	31.67
Efficiency %		44.44	60.60	63.15

$$\text{The length of the warp sheet} = \frac{0.6 \times 1,000}{30} = 20 \text{ km} = 20,000 \text{ m}$$

So running time = 20,000/1,000 = 20 min.

Creeling time = 45 min/creel.

In the case of single end creel, this 45 min will be divided among three warper's beams as from one cone of 2 kg mass at least three beams will be made.

So creeling time allocated to one warper's beam = 15 min.

For duplicated creel, the headstock is moved in front of the new creel, which is ready with full packages, so no creeling time is considered. However, it needs the headstock moving time, that is 3 min.

$$\text{Number of breaks in warping/beam} = \frac{500 \times 20,000 \times 0.1}{100 \times 1,000} = 10$$

Repair time = Number of breaks × repair time per break = 10 × 0.5 = 5 min

Transfer failure = 2% of 500 ends = 10

Time for repairing the transfer failure = 10 × 0.5 = 5 min.

This 5 min should be equally allocated among multiple warper's beams as from one cone of 2 kg mass at least three beams will be made.

So repair time for transfer failure allocated to one warper's beam
= 5/3 min = 1.67 min.

From this example, it can be seen that the type of creel can greatly influence the efficiency of the warping process.

3.4 SECTIONAL WARPING

Sectional warping is preferred over beam warping for multi-coloured warp and speciality warp to minimise the wastage. Here the entire width of the warping drum is not utilised simultaneously. It is developed section by section as depicted in Figure 3.6. As only one section is built at a time, a support is needed at one side of the drum. Otherwise the sections will collapse after winding some layers as the yarns are parallel wound where the angle of wind is 0°. This problem is solved by making one side of the drum inclined. The inclination can be of two types:

1. Fixed angle

2. Variable angle (7°, 9°, 11° etc.)

As the winding of one layer of yarns is completed on the drum, the section of yarns is given a requisite traverse so that the yarn at the extreme

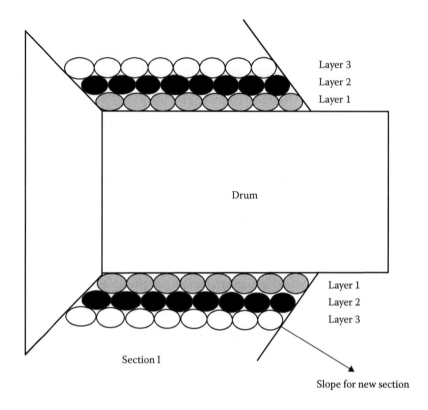

FIGURE 3.6   Schematic representation of sectional warping principle.

left corner touches the inclined surface. Thus it gets support from the inclined surface.

As the process continues, the thickness (or height) of the section gradually increases. When the requisite length has been wound in a section, the next section is started by shifting the expandable reed assembly by suitable distance as indicated by s in Figure 3.7. This distance s is called section width.

If α is the angle of inclination, x is the traverse given to the section and h is the height or thickness of the section, then

$$\frac{h}{x} = \frac{H}{T} = \tan \alpha \tag{3.1}$$

So

$$x = \frac{h}{\tan \alpha} \tag{3.2}$$

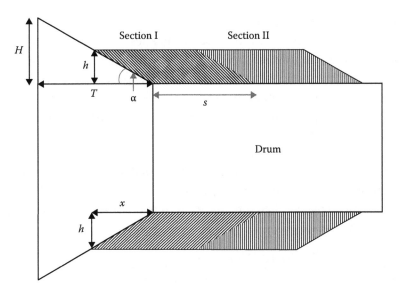

FIGURE 3.7 Traverse in sectional warping.

The volume of yarn contained inside one section can be calculated from the volume of a shaded parallelogram.

The area of a shaded parallelogram $= s \times h = s \times x \tan \alpha$

The average length of warp sheet wound on a drum having empty diameter d and section height h is $= \pi (d + h)$.

Therefore,

The volume of yarn inside one section is $= \pi(d+h)s \times x \tan \alpha$

Alternatively, the volume of yarn in one section can be calculated using the concept of hollow cylinder whose inner and outer diameters are d and $(d + 2h)$, respectively. If the height of the hollow cylinder, that is section width is s, as indicated in Figure 3.7, then the volume can be calculated as follows:

$$\text{Volume} = \frac{\pi\left\{(d+2h)^2 - d^2\right\}}{4} \times s$$

$$= \frac{\pi\left\{(2d+2h) \times 2h\right\}}{4} \times s$$

$$= \pi(d+h) \times h \times s = \pi(d+h)s \times x \tan \alpha \qquad (3.3)$$

For a drum with fixed angle, if the yarn is coarser then one layer of the warp ribbon will result in greater increase in thickness (Δh) and thus to match the inclination, the traverse speed (Δx) should be higher. The machines are designed with provisions to change the traverse speed so that a wide range of yarn count can be managed. For fixed angle drums, only one variable, that is traverse speed can be adjusted while with variable angle drums, the traverse speed and angle of inclination can be varied.

For drums with variable angle, the angle is changed by changing the inclination of metal plates which are supported at the end of the drum. When the angle is increased, larger gaps are created between the neighbouring metal plates. Therefore, the yarns remain unsupported at the gaps between two metal plates.

3.5 STAGES OF WARPING

The flow chart of the warping process is shown in Figure 3.8.

3.5.1 Creel

Creel holds the supply packages which may be cones or cheeses. The yarns pass through the tensioner and stop motions mounted on the creel. The creel capacity of modern warping machines can go up to 700–1000.

3.5.2 Leasing

Leasing is a system by which the position of the yarns is maintained in the warp sheet. Generally it is done by dividing the yarns in two groups (odd and even). If odd numbered yarns are passing over a lease rod, then the even numbered yarns will pass below the rod. The relative positions of the yarns will reverse in the case of second lease rod. Leasing helps the operator to locate the position of broken yarns and thus prevents the occurrence of crossed yarn in the beam. Before winding the warp yarns on the drum of sectional warping, a band or rope is inserted through the groups of odd and even yarns and tied at one side of the drum. In the case of beam warping, this is done at the end when the beam is complete.

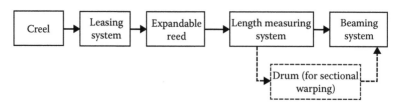

FIGURE 3.8 Sequence of warping process.

3.5.3 Expandable Reed

Expandable reed is used to control the spacing between consecutive yarns. The two limbs of 'V' shaped expandable reed can be expanded or collapsed as per the required spacing of yarns. Figure 3.9 shows the expandable reed and lease rods of a sectional warping machine.

3.5.4 Beaming

Beaming is the process of winding all the yarns simultaneously on the warper's beam. In beam warping, the warper's beam can be driven by two ways:

1. Direct drive or spindle drive

2. Indirect drive or surface drive

In the case of direct drive, the warper's beam is driven by gears. As the diameter of the beam increases, the rotational speed of the beam is reduced in order to keep constant warping speed. In the case of indirect drive, the warper's beam is rotated by frictional contact with another drum. In this case, the rotational speed of the warper's beam reduces automatically as its diameter increases. Thus the warping speed remains constant. Direct drive would be preferable if the yarn is delicate, which can be damaged due to abrasion with driving drum. In modern warping machine (Benninger), the warping speed is around 1200–1400 m/min. The full beam diameter is 1–1.4 m.

In sectional warping, all the sections are simultaneously transferred from the drum to the double flanged warper's beam (Figure 3.10). The drum

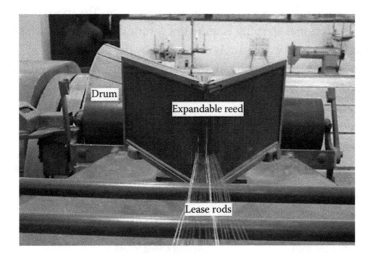

FIGURE 3.9   Expandable reed and lease rods.

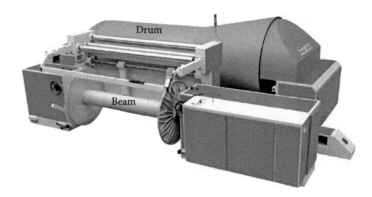

FIGURE 3.10 Sectional warping drum and beam. (Courtesy of Prashant Group of Industries, Ahmedabad, India.)

diameter is quite high in the case of sectional warping so that the change in surface speed of the drum, with the increase in thickness of section, is marginal. For example, if the drum diameter is 0.8 m, then one revolution of the drum will wind 2.5 m warp sheet (section). Therefore, around 250 m warp sheet (section) can be wound with 100 revolutions of the drum, which will increase the effective drum diameter by 200 × flattened yarn diameter. If yarn diameter is 0.2 mm, then increase in effective drum diameter will be 200 × 0.2 mm, that is 4 cm without considering any yarn flattening. Therefore, increase in surface speed of the drum will be around 4/80 or 5%. While transferring all the sections from the drum to the warper's beam, the former is rotated by the torque created by the tension of warp sheet, whereas positive drive is given to the warper's beam. The speed of beaming process in sectional warping is quite low (around 300 m/min). The density of yarns on the warper's beam is controlled by the press rolls.

Table 3.2 presents the comparison between beam warping and sectional warping.

TABLE 3.2 Comparison between Beam Warping and Sectional Warping

Beam Warping	Sectional Warping
Used for high-volume production	Used for small-volume and customised production (stripes and specialised yarns)
One-step process	Two-step process
High creel capacity is required	Low creel capacity is sufficient
Comparatively less expensive	Comparatively more expensive
Beaming speed is high	Beaming speed is low
More common	Less common

3.6 DEVELOPMENTS IN WARPING

Warping machine manufacturers are continuously trying to increase the machine speed and to ensure the best possible quality of the warper's beam. Maintaining uniform density of the warper's beam at different layers is one of the major quality challenges. Benninger has developed an active kick-back system for this purpose as shown in Figure 3.11 (Benninger, 2003). A press roller made up of hard paper is used to exert pressure on the warp layer formed on the warper's beam. The increasing diameter of the warp layer on the beam forces the press roller back against the resistance of the pre-set pressing force. Due to this indirect pressing action, the winding is always uniform and warp layer remains perfectly cylindrical. When the brake is applied, the press roller is moved back immediately, thus avoiding friction between roll and yarn.

In Benninger Bentronic sectional warping machine, the direction of rotation of the drum has been reversed as shown in Figure 3.12. This causes less deflection in the path of the warp sheet.

3.6.1 Sample Warping Machine

Sample warping machine has become very popular in the industry for the quick development of small samples with minimum wastage. Like sectional warping, it is also a two-stage process. The yarn packages are mounted on the creel, which can be of rotating type. The yarns are first wound on the drum which has a big diameter. Unlike sectional warping drum, there is no inclination at the side of the drum. In some of the

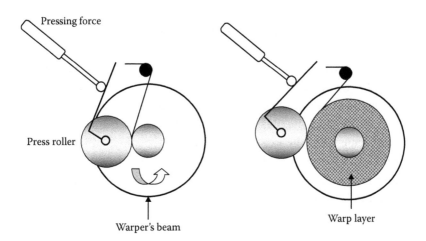

FIGURE 3.11   Active kick-back system.

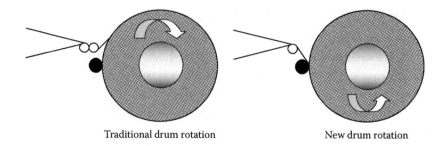

Traditional drum rotation   New drum rotation

FIGURE 3.12   Reverse drum rotation in sectional warping machine.

TABLE 3.3   Similarities and Differences between Sectional and Sample Warping

Similarities	Differences
Both are used for small sample production.	Sample warping can go up to maximum 1500 m length.
Multi-coloured warp sheet can be produced.	Sample warping is possible with only one or few supply packages. This is not possible in the case of sectional warping.
Low level of yarn wastage.	Any colour sequence is possible in sample warping.

models, up to 128 different coloured yarns can be used in warp. After winding the requisite length of warp sheet on the drum, it is transferred to the actual beam. This system can produce warp sheet from a few metres to 1500 m. The similarities and differences between sectional and sample warping are given in Table 3.3.

3.6.2 Smart Beam

Carl Mayer has introduced the concept of smart beam in warping. All the data relevant to the warper's beam is transferred from the computer of warping machine to an RFID (radio frequency identification) chip attached to the flange. When the warper's beam is placed on the creel of the sizing machine, all the data are automatically transferred to the computer of sizing machine. Based on the supplied data, the system proposes a sizing recipe from the existing database. If the supplied information of warp does not match with the existing database, then the system suggests a sizing recipe for consideration by the sizing operator. The following information is stored in the RFID chip:

- Style and order number
- Beam number

- Fibre blend

- Yarn count and number of ply

- End density

- Warp length

- Number of layers of warp

- List and location of missing warp yarns

NUMERICAL PROBLEMS

3.1 A full beam produced on a direct warping system has a 2 m width and contains 500 ends of 30 tex yarn. The empty and full beam diameters are 50 and 75 cm, respectively. If the beam density is 0.4 g/cm³, then calculate the length of warp and its mass in kg.

Solution:

Beam width (L) = 2 m = 200 cm
Number of ends = 500
Count = 30 tex
Empty beam diameter (d) = 50 cm
Full beam diameter (D) = 75 cm

So

$$\text{Volume of yarn on the beam} = \frac{\pi}{4}\left(D^2 - d^2\right) \times L$$

$$= \frac{\pi}{4}\left(75^2 - 50^2\right) \times 200 \text{ cm}^3$$

$$= 490,625 \text{ cm}^3$$

So

$$\text{Mass of yarn on the beam} = \text{Volume} \times \text{Density}$$

$$= 490,625 \times 0.4 = 196,250 \text{ g}$$

$$= 196.25 \text{ kg}$$

Therefore,

$$\text{Mass of single yarn} = \frac{196,250}{500} = 392.5 \text{ g}$$

So

$$\text{Length of warp} = \frac{392.5}{30} \text{km} = 13.08 \text{ km}$$

So the total mass of yarn on the beam is 196.25 kg and the length of warp sheet is 13.08 km.

3.2 A multi-coloured warp of 20 tex is to be wound on a sectional warping drum having 1 m diameter and 15° inclination at the side. The warp sheet is 1.5 m wide and contains 6480 ends. The density of the material on the beam is 500 kg/m³. The maximum thickness of yarn layers on the beam could be 10 cm. Determine the length of warp and the traverse length per section.

Solution:
Figure 3.13 shows the geometry of sectional warping drum.

Yarn count = 20 tex
Warp width (L) = 1.5 m
Empty beam diameter (d) = 1 m
Full diameter of beam (D) = $d + 2h$

where h is the maximum thickness of yarn layers on the beam = 10 cm = 0.1 m.
So

$$D = (1 + 2 \times 0.1) \text{m} = 1.2 \text{ m}$$

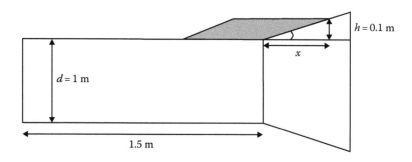

FIGURE 3.13   Geometry of sectional warping drum.

$$\text{Volume of the yarn} = \frac{\pi \times (D^2 - d^2) \times L}{4} \, m^3$$

$$= \frac{\pi \times (1.2^2 - 1^2) \times 1.5}{4} \, m^3$$

$$= 0.5181 \, m^3$$

Mass of the yarn = 0.5181 × 500 kg = 259.05 kg

$$\text{Mass of a single end} = \frac{259.05 \times 1000}{6480} = 40 \, g$$

So

$$\text{Length of the warp} = \frac{40}{tex} \, km = \frac{40}{20} \, km = 2000 \, m$$

Here

$$\frac{h}{x} = \tan 15°$$

where x is the traverse length per section.

Or

$$\frac{10}{x} = \tan 15°$$

or

$$x = 37.33 \, cm$$

So traverse length per section is 32.33 cm.

So the length of warp and traverse length per section are 2000 m and 32.33 cm, respectively.

3.3 The drum diameter of a sectional warping machine is 1 m. The maximum permissible thickness of warp layer is 5 cm. If the yarn (100% cotton) count is 20 Ne and inclination of the drum at the sides is 10°, then calculate the length of the warp sheet. If the lateral traverse is given to the yarn section after every revolution of the drum, then calculate the lateral traverse per drum revolution.

Solution:

$$\text{Drum diameter} = d = 100 \, cm.$$

$$\text{Thickness of warp layer} = h = 5 \, cm.$$

$$\text{Yarn diameter} = \frac{2.54}{28\sqrt{20}} = 0.02 \, cm.$$

Therefore,

$$\text{The number of yarn layers} = \frac{5}{0.02} = 250$$

$$\text{Length of warp sheet} = \pi(d+h) \times \text{number of yarn layers}$$

$$= \pi(100+5) \times 250 = 82{,}425 \text{ cm}$$

$$= 824.25 \text{ m.}$$

The increase of thickness of yarn layer after one revolution of drum is Δh = yarn diameter.
Lateral traverse given to the yarn section after every revolution $= \Delta x$

Now,

$$\frac{\Delta h}{\Delta x} = \tan 10°$$

Or

$$\Delta x = \frac{\Delta h}{\tan 10°} = \frac{0.02}{0.1763} = 0.113 \text{ cm.}$$

So length of the warp sheet is 824.25 m and lateral traverse after every revolution of drum is 0.113 cm.

REFERENCES

Banerjee, P. K. 2015. *Principles of Fabric Formation*. Boca Raton, FL: CRC Press.

Booth, J. E. 1977. *Textile Mathematics*, Vol. III. Manchester, UK: The Textile Institute.

Lord, P. R. and Mohamed, M. H. 1982. *Weaving: Conversion of Yarn to Fabric*, 2nd edn. Merrow, UK: Merrow Technical Library.

Technical literature of warping machine Ben-direct, 2003, Benninger Co. Ltd., Uzwil, Switzerland.

CHAPTER 4

Warp Sizing

4.1 OBJECTIVES

The objective of warp sizing is to improve the weaveability of yarns by applying a uniform protective coating on the yarn surface. The protruding hairs are also laid on the yarn surface due to sizing.

During the weaving process, the warp yarns are subjected to abrasion with various loom components, such as backrest, heald eyes, reed, front rest and shuttle. Moreover, in shedding operation, warp yarns abrade against each other. Size coating protects the yarn structure from abrasion. Therefore, the warp breakage rate in the loom reduces. The benefits of sizing are as follows.

- It prevents the warp yarn breakage due to abrasion between neighbouring yarns or with weaving machine components.

- It improves the yarn strength by 10%–20%, although it is not the primary objective of sizing process.

4.2 CHARACTERISTICS OF SIZED YARN

The sized yarns demonstrate the following behaviour in comparison to unsized yarns:

- Higher strength

- Lower elongation

- Higher bending rigidity

- Higher abrasion resistance

- Lower hairiness

- Lower frictional resistance

Sized yarns have higher tenacity than unsized yarn as fibre-to-fibre slippage is reduced after sizing. Therefore, more proportion of fibres breaks in sized yarn during tensile failure, contributing towards yarn strength. As the mobility and slippage of fibres get restricted after sizing, the elongation reduces. The bending rigidity also increases after sizing as a large number of fibres behave as a bundle due to the presence of a matrix in the form of size materials. The abrasion resistance increases as the size film protects the yarn structure during abrasion. The protruding hairs get embedded on the yarn body due to the film formed by the sizing material.

4.3 IMPORTANT DEFINITIONS OF SIZING

'Size paste concentration', 'size pick-up' and 'size add-on' are some of the terms used frequently in the discussion of sizing process. They are expressed as follows:

$$\text{Size paste concentration (\%)} = \frac{\text{Oven dry mass of size materials}}{\text{Mass of size paste}} \times 100$$

$$= \frac{S}{P} \times 100$$

$$\text{Size add-on (\%)} = \frac{\text{Oven dry mass of size materials in yarns}}{\text{Oven dry mass of unsized yarns}} \times 100 = \frac{S}{Y} \times 100$$

$$\text{Wet pick-up} = \frac{\text{Mass of size paste in yarns}}{\text{Oven dry mass of unsized yarns}} = \frac{P}{Y}$$

$$= \left(\frac{S}{Y} \times 100 \right) \times \left(\frac{P}{S} \times \frac{1}{100} \right)$$

$$= \frac{\text{Add-on \%}}{\text{Concentration \%}}$$

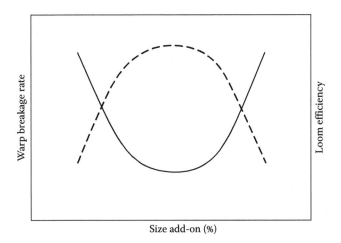

FIGURE 4.1   Sizing-weaving curve.

4.4 SIZING-WEAVING CURVE

In sizing, depending on the size materials used, there is a target add-on for the optimum performance of the warp yarns during weaving. This can be understood from the sizing-weaving curve (Figure 4.1). The solid line represents the warp breakage rate, whereas the broken line implies loom efficiency. At very low levels of size add-on, the yarn is not adequately covered by the size film and therefore the yarn is not fully protected from the abrasion with various loom parts. So warp breakage rate is generally high at very low level of size add-on. The performance of the yarn in weaving improves as the size add-on increases. The optimum add-on level is marked by very low level of warp breakage rate. The sizing-weaving curve has a plateau region when the optimum size add-on is achieved. However, if the size add-on is higher than the optimum level, then warp breakage rate increases again largely due to the loss of elongation and increase in stiffness, that is bending rigidity of the yarns.

In general, the range of size add-on is 8%–14% for ring spun cotton yarns and 4%–6% for synthetic filament yarns.

4.5 SIZE ENCAPSULATION AND SIZE PENETRATION

Although add-on primarily influences the weaving performance, it is possible to have different weaving performances even at the same level of size add-on. This can happen due to differences in (1) size coating or encapsulation and (2) size penetration. This is explained in Figure 4.2.

  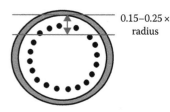

0.15–0.25 × radius

FIGURE 4.2 Size coating or encapsulation, size penetration and optimum coating and penetration.

In the first case, the size material has formed a uniform coating on the yarn surface. However, the penetration of the size material inside the yarn structure is inadequate. This may provide good protective coating on yarn and also tackle the hairiness problem. However, the adhesion of the size film with the yarn will be poor. The size coating will also be very stiff. Therefore, during the abrasion with various machine parts and neighbouring yarns, the size will fall (shed) and thus the weaving performance of the yarn will be poor.

In the second case, the coating or encapsulation on the yarn surface is inadequate. Therefore, the yarn will not get adequate protection against the abrasion. Besides, the size materials have penetrated too much inside the yarn, which is not desirable.

The third case is the optimum one where there is a thin but uniform coating of size materials on the yarn surface. Moreover, the size materials have also penetrated to some extent in the yarn structure, which ensures good adhesion between the yarn and the size coating. For optimum weaving performance, size should penetrate to a distance equal to 15%–25% of yarn radius.

4.6 SIZING MATERIALS

The choice of sizing materials, optimum level of size add-on and sizing parameters depend on the following factors:

- Type of fibre
- Type of sizing materials
- Yarn spinning technology
- Yarn count and twist
- Level of yarn hairiness

- Fabric sett and weave

- Loom type and loom speed

Table 4.1 shows a non-exhaustive list of size materials used for various fibres (Goswami et al., 2004).

Rotor spun yarns are bulkier than equivalent ring spun yarn. On the other hand, compact spun yarns have higher packing factor than the equivalent ring spun yarns. Therefore, higher and lower viscosities of size paste should be used for rotor and compact spun yarns, respectively, to control the level of penetration of size materials inside the yarn structure.

Finer yarns generally require higher add-on as they have higher specific surface area and thus more size materials are needed to form uniform coating on the yarn surface. Higher twist reduces the yarn hairiness and therefore add-on requirement reduces. Higher twist increases the packing factor of the yarn and thus lower size paste viscosity may be required to ensure optimum penetration of size materials. Compact spun yarns require lower size add-on than equivalent ring spun yarns as the former has lesser hairiness.

Higher fabric sett (ends per cm and picks per cm) necessitates increased add-on. Higher end density causes more abrasion among the closely placed warp yarns and thus more encapsulation is required. Higher picks per inch implies lower fabric take-up after the insertion of each pick. Thus a segment of warp yarns is subjected to more number of abrasion cycles with various components of loom such as reed dents and heald eyes. Therefore, the add-on requirement increases. Plain weave has maximum number of interlacement and all the ends change their positions from the top shed line to the bottom shed line or vice versa after each pick. So the abrasion between the warp yarns is very high. In the case of a two up one down (2/1) twill weave, one-third of the ends do not change their position between consecutive picks and thus the abrasion among the yarns is lower as compared to that of plain weave. Therefore, plain weave should

TABLE 4.1 Size Materials for Various Fibres

Fibre	Size Material
Cotton	Starch, PVA (fully hydrolysed)
Polyester	PVA (partially hydrolysed), poly-acrylates, poly-vinyl acetate
Viscose	PVA (fully hydrolysed)
Acetate	PVA (partially hydrolysed)
Nylon	Poly-acrylic acid, PVA (partially hydrolysed)

require more add-on than other weaves. Size materials to be used in water-jet loom have a special requirement. The size should be insoluble in water once it has been dried after application on warp.

4.6.1 Desirable Properties of Sizing Materials

The sizing material must fulfil some essential properties and at the same time it is expected that it will have some additional desirable properties. The sizing material must form a smooth and uniform coating on the yarn surface. This is known as film forming property. The coating will not only embed the protruding fibres (hairs) on the yarn body but also protect the yarn structure from repeated abrasion during weaving. The size film should adhere to the fibres strongly to prevent shedding (dropping of size film). The film should also have enough flexibility to cope with the flexing or bending of yarns around the back rest, heald eyes and other loom components. Sizing material should be easily removable during desizing and should create minimum environmental pollution. The non-exhaustive list of essential and desirable properties of sizing materials is given in Table 4.2.

4.6.2 Composition of Sizing Material

The specific composition of sizing material depends on the fibre type, yarn type, yarn count, fabric sett and so on. However, the materials can be classified under the categories of adhesive, softening agent, antimicrobial agent and so on. The adhesive part is responsible for forming a protective film and adhering to the fibres. Softening agents make the film flexible so that the film can bend easily without forming cracks. Antimicrobial agents are added to thwart the mildew growth on the size film. Sometimes, weighting agents and dyes are also added to fulfil specific requirements.

Cotton yarns, in general, are sized by starch which forms the adhesive component of the size mix. The reason behind the popularity of starch can be attributed to the following factors.

TABLE 4.2   Essential and Desirable Properties of Sizing Materials

Film forming	Controllable viscosity
Adhesion	Easy removal and recyclability
Film flexibility and elasticity	Neutral pH
Lubrication	Non-polluting
Bacterial resistance	Cheap

- Starch is chemically same with cotton and rayon and thus the adhesion is very good.

- Desizing of starch is easy.

- It is relatively cheap.

- The properties can be tuned to cope with the need.

However, starch gives a very stiff film. It has higher biological oxygen demand (BOD). Cooking of starch is required to attain uniformity. Besides, starch has poor bacterial resistance. Overall, the positive attributes of starch dominate over the drawbacks and thus it is still being used in the industry as the primary materials for the sizing of cotton yarns.

The softening materials compensate for the abrasive and harsh feel that is provided by most of the starches. Softeners also lubricate the yarns so that they can pass easily over machine parts without shedding. They also prevent the sticking of size ingredients over the drying cylinders. Mutton tallow, which is composed of glycerides of palmitic, stearic and oleic acids, can be used as softener. The proportion of softener in the size mix is very crucial as excessive use of it deteriorates the strength of size film.

4.6.3 Starch

Starches are available from the seed, root or pith of plants. Corn, rice and wheat are the examples of seed starch. Potato and tapioca starches are obtained from roots. Sago starch is obtained from pith. Starches are prepared by grinding the seed, root or pith into fine flour. When the flour is mixed with water and cooked, it produces a thick and smooth glutinous solution. Corn (maize) starch is the most popular type of starch used in textile sizing. Around 50% of the corn is composed of starch. Corn starch is generally preferred for the sizing of coarse and medium count yarns. Potato yields around 20% starch. It is slow congealing type and therefore gets more chance to penetrate within the yarn structure. It forms a smooth and pliable film on the yarn body. Potato starch is preferred for sizing finer yarns.

4.6.3.1 Chemical Structure of Starch

Cellulose and starch are chemically same and both are polymers of glucose (Goswami et al., 2004). Glucose can have two structural (anomeric) forms known as α and β (Figure 4.3).

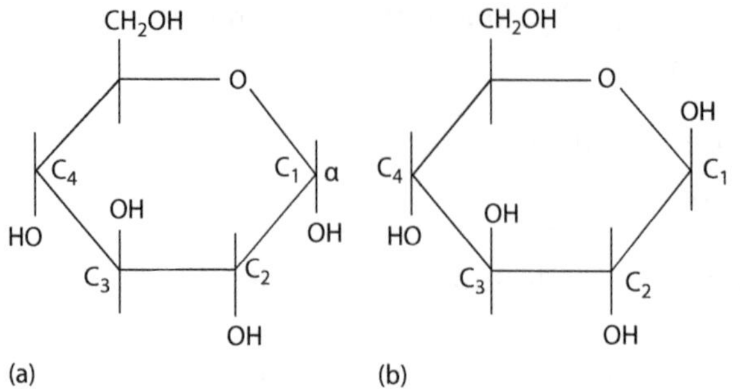

FIGURE 4.3   Structure of (a) α-glucose and (b) β-glucose.

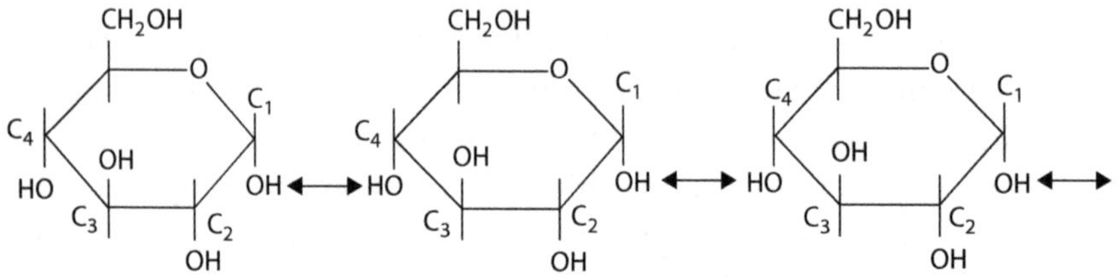

FIGURE 4.4   Condensation of α-glucose units.

It can be seen from Figure 4.3 that the position of −OH group in the pyranose ring at C_2, C_3 and C_4 are same in α and β glucose. For α-glucose, which is the building block of starch, both the −OH groups are down with respect to the next glucose molecule. Therefore, no change in orientation is needed for condensation as shown in Figure 4.4.

On the other hand, in β-glucose, which is the building block of cellulose, the −OH group is up at C_1 and down at C_4. As a result, every other glucose unit must flip over before the 1,4 hydroxyl groups can come closer for condensation as shown in Figure 4.5.

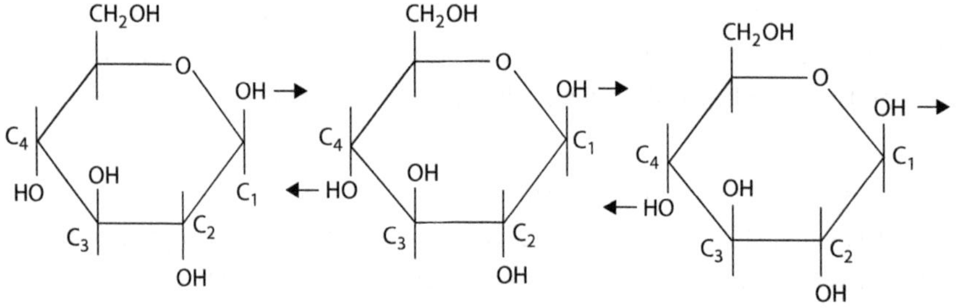

FIGURE 4.5   Condensation of β-glucose units.

TABLE 4.3 Differences between Amylose and Amylopectin

Amylose	Amylopectin
Provides strength	Prevents rapid gelling
Low molecular weight	Relatively high molecular weight
Water soluble	Water insoluble
Proportion in starch: 20%–30%	Proportion in starch: 70%–80%

Starches have two components. The straight chain component is called amylose. The branched chain component is called amylopectin. The differences between amylose and amylopectin are highlighted in Table 4.3.

The proportion of amylose and amylopectin differs depending on the source of starch. For example, in potato starch the ratio is 20:80, whereas in wheat starch the ratio is 25:75.

4.6.3.2 Cooking of Starch

Starch remains tightly bound in granules and therefore it does not act as adhesive in cold water. Cooking of starch is required to make it soluble in water. The change in viscosity of starch solution during cooking is shown in Figure 4.6.

Within the granule, the chain molecules of amylose and amylopectin are arranged radially in stratified layers. External heat energy is required

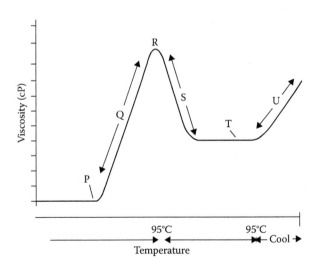

FIGURE 4.6 Cooking of starch.

for the penetration of water molecule within the structure. The temperature at which the thermal energy becomes sufficient to overcome hydrogen bonding within the structure is called 'gelatinisation' temperature (P). Crystallization of starch is lost during gelation. As water penetrates, the chain molecules are pushed away from each other, causing swelling of the starch granule. This is marked by increase in viscosity of the solution (Q). This continues up to the point R. Aided by the continuous shearing provided by the stirring, the starch granules finally break. The chain molecules of amylose and amylopectin come out in the solution, causing reduction in viscosity (S). When all the granules burst, the viscosity stabilizes or levels off (T). When the solution is cooled, the starch gels due to the formation of a rigid interlocked micelle-like structure having hydrogen bonding (U). This gel form of starch can create a continuous coating on the yarn surface.

4.6.3.3 Acid Treatment of Starch

The viscosity of the sizing paste influences the wet pick-up and resultant add-on %. The viscosity is influenced by the concentration of starch (solid content) and molecular chain length of starch. To reduce the concentration of size paste, by keeping the solid content same, acid treatment is performed. Aqueous solution of starch is treated with hydrochloric acids at specified temperature and duration. The acid cleaves the polymer at the glycoside linkage and thus the length of the polymer chain is reduced (Figure 4.7). Hence, the viscosity is reduced and the fluidity, which is the reciprocal of viscosity, increases (Banerjee, 2015). The acid-treated starch is often termed 'thin boiling starch' as it results in lower viscosity than the normal starch at a given concentration.

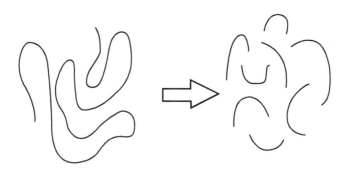

FIGURE 4.7   Acid treatment of starch to reduce the molecular chain length.

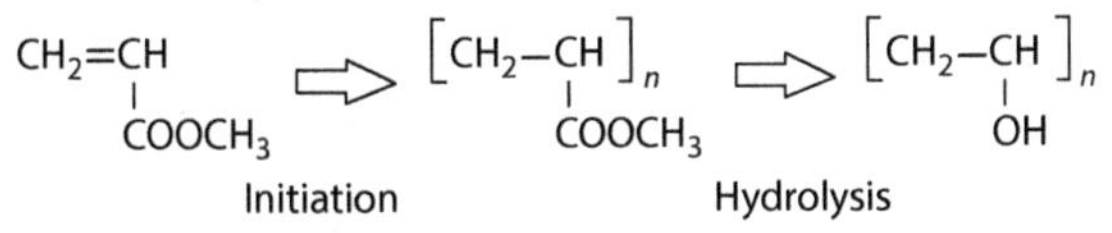

$$CH_2=CH \quad \Rightarrow \quad \left[CH_2-CH \right]_n \quad \Rightarrow \quad \left[CH_2-CH \right]_n$$
$$\qquad\ |\qquad\qquad\qquad\quad |\qquad\qquad\qquad\qquad |$$
$$\qquad COOCH_3 \qquad\qquad COOCH_3 \qquad\qquad\qquad OH$$

Initiation　　　　　　Hydrolysis

FIGURE 4.8   Steps for manufacturing PVA.

4.6.4 Polyvinyl Alcohol (PVA)

Polyvinyl alcohol (PVA) is a very versatile sizing material. It can be used for sizing cotton, rayon, polyester and their blends. It is manufactured by polymerizing vinyl acetate monomers (VAM) to polyvinyl acetate. In the second step, acetate ($COOCH_3$) groups are substituted with hydroxyl groups ($-OH$) by hydrolysis (Figure 4.8).

4.6.4.1 Degree of Hydrolysis

The properties of PVA are largely governed by the degree of hydrolysis. Table 4.4 presents the degree of hydrolysis and application range for various grades of PVA.

If the degree of hydrolysis is maximum (>99%), then it is called super-hydrolysed PVA. The formation of hydrogen bonding becomes very intense in this case (Figure 4.9) and thus the strength of super-hydrolysed PVA film is very high (Figure 4.10). Besides, its solubility in water is lower as compared to that of partially hydrolysed PVA. Thus the desizing becomes difficult and higher temperature is required for desizing of super-hydrolysed PVA. Therefore, this grade is generally not preferred for sizing.

Partially hydrolysed PVA exhibits lower film strength. The presence of a big functional group ($COOCH_3$) in the side chain reduces the strength of partially hydrolysed PVA film. Thus it provides the advantages of easy splitting (separation) of yarns after drying and less disruption of size film. Besides, partially hydrolysed PVA shows better adhesion to the hydrophobic fibres than the fully hydrolysed PVA (Table 4.5).

Solubility in water is also better for the partially hydrolysed PVA. In partially hydrolysed PVA, the acetate group, due to its bigger size, acts

TABLE 4.4   Degree of Hydrolysis in Different Grades of PVA

PVA Grade	Degree of Hydrolysis (%)	Application
Super hydrolysed	>99	Not a preferred material for sizing
Fully hydrolysed	98–99	100% Cotton
Intermediate hydrolysed	95–98	Polyester and other synthetic fibres and blends
Partially hydrolysed	87–90	Polyester and other synthetic fibres and blends

Source: Technical literature of Selvol™, 2011.

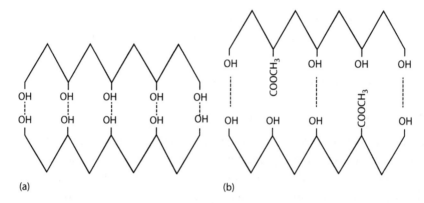

FIGURE 4.9 Hydrogen bonding in (a) super and (b) partially hydrolysed PVA.

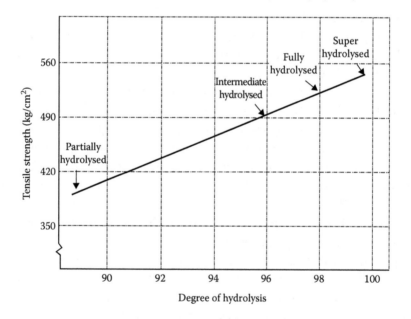

FIGURE 4.10 Effect of hydrolysis on the strength of PVA film.

TABLE 4.5 Adhesion of PVA with Synthetic Fibres

Type of Fibre	Adhesion Strength (g/mm²)	
	Partially Hydrolysed	**Fully Hydrolysed**
Acetate	10	2
Nylon	11	6
Acrylic	9	4
Polyester	7	1

Source: Technical literature of Selvol™, 2011.

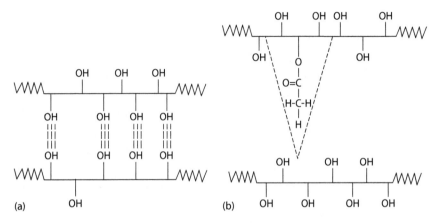

FIGURE 4.11 Molecular chains of (a) fully hydrolysed and (b) partially hydrolysed PVA.

as an obstacle against the close packing of molecular chain (Figure 4.11). Therefore, the penetration of water molecule is rather easy in the case of partially hydrolysed PVA, thus making the desizing operation facile.

Figure 4.12 depicts the abrasion resistance of yarns sized with partially, intermediate and fully hydrolysed PVA. The former provides better waving performance in terms of the following:

- Lower shedding or dropping of size

- Lower yarn hairiness

- Lower size add-on for same level of warp breakage rate

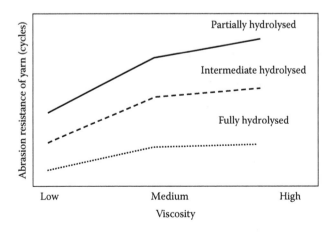

FIGURE 4.12 Abrasion resistance of yarns sized with PVA.

4.6.4.2 Degree of Polymerisation and Viscosity

PVA is available in different viscosity ranges, based on its degree of polymerization (DP). Higher DP leads to greater molecular weight and higher viscosity. The viscosity ranges from 5 to 65 cP. For most of the spun yarns, medium viscosity is preferred. If the viscosity is very low, then size paste penetrates too much within the yarn structure affecting the flexibility and extensibility. On the other hand, if the viscosity is too high, then size film forms over the yarn surface without much adhesion to the fibres. These surface coatings of size materials are prone to dropping. To maintain the balance, a mixture of low viscosity and high viscosity PVA is often used.

4.6.5 Typical Recipe of Sizing

4.6.5.1 Carded Cotton Yarn

Modified starch: 10.5% on the weight of water

PVA: 2.86%

Acrylic binder (liquid): 6.6%

Lubricant: 0.7%.

Paste viscosity: 6.5 ± 0.2 s, Solid content: 12%–13%

4.6.5.2 Combed Cotton Yarn

Modified starch: 12.5% on the weight of water

PVA: 3.0%

Acrylic binder (liquid): 6.91%

Lubricant: 0.87%

Paste viscosity: 6.4 ± 0.2 s, Solid content: 12%

4.6.5.3 Polyester-Cotton Blended Yarn

Modified Starch: 13% on the weight of water

PVA: 3.6%

Acrylic binder (liquid): 8.4%

Lubricant: 0.87%

Antistatic: 0.5 kg

Paste viscosity: 7 ± 0.2 s, Solid content: 12%–13%

4.6.6 Steps for Preparing the Size Paste

The typical steps followed in industry for preparing the size paste is as follows:

1. Take standard volume (700 L) of water (normal temperature) into the pre-mixture through water flow meter and start the stirrer.

2. Properly weigh all the chemicals required.

3. Add modified starch, PVA and then acrylic binder slowly into the pre-mixture vessel. Stir the mixture for 15 min.

4. Start the stirrer of the cooker and then transfer the mixture to the closed cooker from the pre-mixture vessel. Close all the open valves of the cooker.

5. Heat the cooker mixture with the injection of direct steam. The steam supply will cut automatically when the cooker temperature reaches the preset value (110°C). It takes around 30 min to attain the temperature.

6. Cook the mixture for another 40 min and then the temperature of the cooker will reach around 120°C–125°C depending on the size recipe.

7. Start the stirrer of the storage vessel and then transfer the cooked size paste to the storage vessel.

8. Add lubricant with the size paste in the storage vessel.

9. Maintain the storage temperature at around 80°C–90°C.

10. The paste is now ready to get transferred to the sow box of sizing machine.

4.7 SIZING MACHINE

The sizing machine can be divided into four main zones as shown in Figure 4.13. The zones are as follows:

1. Creel zone

2. Size box zone

3. Drying zone

4. Headstock zone

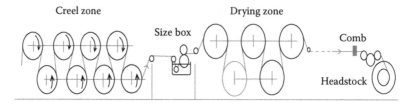

FIGURE 4.13   Zones of a sizing machine.

The creel zone contains a large number of warper's beams which can be arranged in different fashion depending on the design of the creel. Individual warp sheet emerging from the warper's beams are merged together to form the final warp sheet which passes through the size box. During the passage through the size box, the warp sheet picks up size paste and holds a part of the paste after squeezing. Then the wet warp sheet passes through the drying zone and the size paste gets dried. Finally, the warp sheet is wound on the weaver's beams in the headstock zone.

4.7.1 Creel Zone

The creel zone of a sizing machine can have the following types of design:

- Over and under creel (Figures 4.14 and 4.15)

- Equi-tension creel (Figure 4.16)

- Vertical creel (Figure 4.17)

- Inclined creel

In the case of over and under creel (Figure 4.14), the warper's beams are arranged in two rows, having different heights, in an alternate manner.

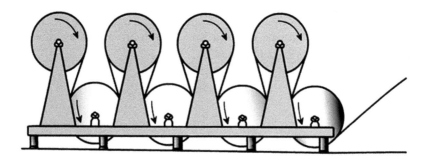

FIGURE 4.14   Over and under creel.

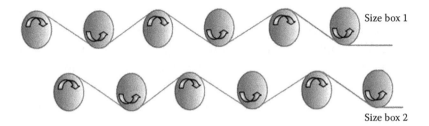

Size box 1

Size box 2

FIGURE 4.15   Over and under creel for two size boxes.

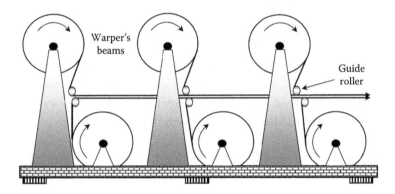

Warper's beams

Guide roller

FIGURE 4.16   Equi-tension creel.

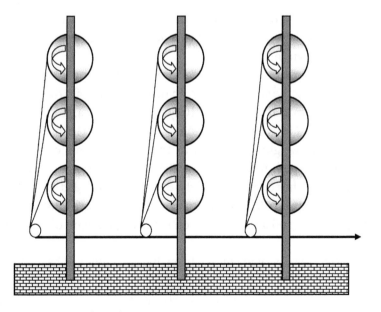

FIGURE 4.17   Vertical creel.

The warp sheet coming out from the rearmost beam passes under the second beam and over the third beam and so on. The individual warp sheets coming out from the beams are merged together to form the final warp sheet. The warp sheet coming from the rearmost beam definitely experiences more tension and stretch than the warp sheet coming from the beam located nearest to the size box. The problem is partially mitigated when two creels are used, one for each of the two size boxes as shown in Figure 4.15. If there are 12 warper's beams, then six beams are mounted on creel one and the remaining six beams are mounted on creel two, reducing the over and under movement of the warp sheet.

In the case of equi-tension creel (Figure 4.16) the pattern of movement of the warp sheet is completely different than that of over and under creel. In equi-tension creel, the warp sheet does not move over and under the beams. A small guide roller, provided with every beam, deflects the warp sheet towards the proper path. Here, the warp sheets are subjected to equal tension and stretch irrespective of the position of the warper's beam.

Another improvement in this direction has been implemented in the inclined creel. Here the height of the beam changes based on its position so that a constant inclination can be maintained in the path of the warp sheet.

All the designs, which have been discussed thus far, require a considerable amount of floor space. The space requirement can be drastically reduced if vertical creels are used. In vertical creels, the beams are stacked vertically as shown in Figure 4.17.

It is very important to maintain adequate and uniform tension in the warp sheet during the sizing process. In older versions of sizing machines, the warper's beam may not be positively driven. The beams are rotated by the torque created by the warp tension (T) against the frictional resistance created by dead weights. However, as the sizing process continues, the radius of the warper's beam reduces. Therefore, the warp tension should be adjusted by either adjusting the dead weight suspended with the rope passing over the two ruffles of the warper's beam (Figure 4.18) or controlling the pneumatic pressure applied on the bearing region of the warper's beams.

From the balance of moment, the following equations can be formulated:

$$T \times R = 2T_2 \times r(1 - e^{-\mu\theta}) \tag{4.1}$$

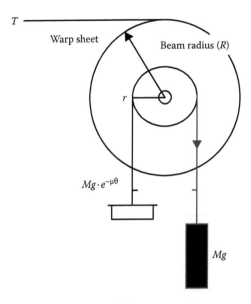

T

Warp sheet

Beam radius (R)

r

$Mg \cdot e^{-\mu\theta}$

Mg

FIGURE 4.18   Warp tension control by a dead weight system.

or

$$T \times R = 2 \times Mg \times r(1 - e^{-\mu\theta}) \qquad (4.2)$$

where
   T is the warp tension
   R is the instantaneous radius of warper's beam
   T_2 is the tension in the tight side of the rope
   r is the ruffle radius
   μ is the coefficient of friction between the rope and the ruffle
   θ is the angle of wrap (in radian) of the rope over the ruffle
   M is the mass of suspended element
   g is acceleration due to gravity

In sizing process, the allowable stretch is 1%–1.5% for cotton and polyester yarns. The stretch can be higher (3%–5%) for viscose and acrylic yarns.

4.7.2 Size Box Zone

This is the zone where the warp sheet is immersed into the size paste and then squeezed under high pressure so that uniform coating of size film is formed over the yarn surface. The process of immersion is called 'dip'

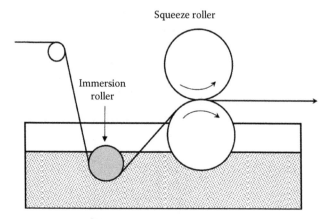

FIGURE 4.19   One dip one nip size box.

and the process of squeezing by means of a pair of squeezing rollers is called 'nip'. The size box can have different number combinations of 'dip' and 'nip' to meet the requirement of various yarns. For filament yarns 'one dip one nip' is preferred (Figure 4.19) as a low add-on is required. For spun yarns made from staple fibres 'two dip two nip' is advisable (Figure 4.20).

Two dip and two nip process allows grater time for immersion of yarns within the size paste and thus this process forms more uniform coating of size film. When the yarns are squeezed by the first pair of squeeze rollers, the yarns get compressed. When the yarns come out of the nip of squeezing rollers, they try to regain their original configuration and therefore an

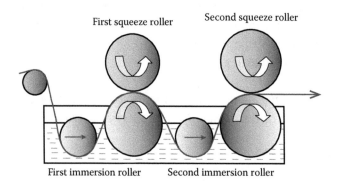

FIGURE 4.20   Two dip two nip size box.

inward pressure is created, which causes more penetration of size materials within the yarn structure.

The size add-on and performance of warp in subsequent operations are influenced by the following parameters related to the size box zone.

4.7.2.1 Viscosity of Size Paste

Viscosity of a fluid indicates its resistance against the flow. The viscosity of size paste is mainly influenced by the concentration (solid content) and temperature of size paste. Higher concentration produces higher viscosity. Viscosity of size paste reduces with the increase in temperature. The wet pick-up generally increases with the increase in viscosity. It has already been discussed that viscosity influences the penetration of size paste within the yarn structure. If more penetration is desired then viscosity should be lowered and vice versa. For bulky yarns, penetration is relatively easy and therefore higher viscosity may be preferred.

Viscosity of size paste can be measured by a Zahn cup, which is a stainless steel cup with a small hole at the centre of the bottom and a long handle attached to the sides of the cup. There are five cup specifications, labelled Zahn cup N#, where N is the number from 1 to 5 as given in Table 4.6. Large number cup sizes are used when the viscosity is high and vice versa. To determine the viscosity, the cup is dipped and completely filled with the size paste. After lifting the cup out of the paste, the user measures the 'efflux time' until the paste streaming out of it breaks up. Viscosity of the paste is calculated from the efflux time using standard formulae as given as follows:

$$\text{Cup \#1:} \quad v = 1.1(t - 29)$$

$$\text{Cup \#2:} \quad v = 3.5(t - 14)$$

$$\text{Cup \#3:} \quad v = 11.7(t - 7.5)$$

$$\text{Cup \#4:} \quad v = 14.8(t - 5)$$

$$\text{Cup \#5:} \quad v = 23t$$

where
t is the efflux time
v is the kinematic viscosity in centistokes

TABLE 4.6   Specifications of Zahn Cups

Model	Volume of Cup (mL)	Diameter of Hole (mm)	Range of Kinematic Viscosity (Centistokes)
Zahn1#	44	1.93	5–60
Zahn2#	44	2.69	20–250
Zahn3#	44	3.86	100–800
Zahn4#	44	4.39	200–1,200
Zahn5#	44	5.41	400–18,000

4.7.2.2 Squeezing Pressure

The squeeze pressure forces out the excess paste picked up by the warp sheet. Besides, the pressure distributes the paste uniformly over the yarn surface and causes the penetration of size paste within the yarn structure. Higher squeeze pressure reduces the wet pick-up and add-on % as shown in Figure 4.21 (Maatoug et al., 2007).

The effect of high-pressure squeezing during sizing was investigated by Hari et al. (1989). It was found that for the same level of size add-on %, the high-pressure squeezing facilitates better penetration of size within the yarn structure. However, the thickness of encapsulation outside the yarn periphery reduces at high-pressure squeezing. This reduces the dropping of size during weaving. The comparison of size encapsulation and penetration at high and low squeezing pressures is presented in Table 4.7.

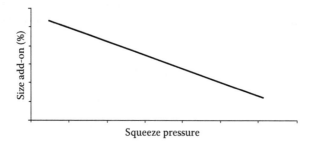

FIGURE 4.21   Effect of squeeze pressure on size add-on %.

TABLE 4.7   Comparison between High- and Low-Pressure Squeezing

Pressure	Size Encapsulation (Film Thickness)	Size Penetration
High	Low	High
Low	High	Low

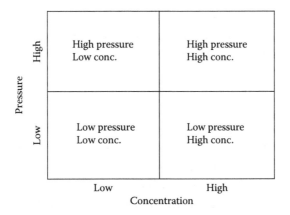

FIGURE 4.22   Different combinations of squeezing pressure and concentration.

Though there was no significant difference in the tensile properties of yarns sized using high and low pressures, the weaveability of the former was much better.

Modern sizing practice recommends the use of high concentration (which results in high viscosity) of size paste and high squeezing pressure (Figure 4.22).

- High pressure and high viscosity combination is preferred as high pressure reduces the wet pick-up and high concentration ensures that targeted add-on will be attained with minimum water evaporation.

- Hypothetically, low pressure and low viscosity combination can also give the same level of wet pick-up as obtained with high pressure and high viscosity combination. But the same level of add-on cannot be obtained after drying due to low concentration of size paste.

- High pressure and low concentration will give very low wet pick-up. On the other hand, low pressure and high concentrations will give very high wet pick-up.

Let us consider one hypothetical example. The targeted size add-on is 10% and oven dry mass of supply warp sheet is 100 kg. If size paste concentration is 20%, then high pressure can be used to achieve wet pick-up of 50 kg. Then in the drying section, 40 kg water will be evaporated to get the target add-on of 10%. In contrast, if the concentration is 10% then low pressure can be used to achieve wet pick-up of 50 kg. Then in the drying section, 45 kg water has to be evaporated to get the add-on of 5%. The second option, that is low pressure and low concentration combination, obviously requires more energy consumption during drying.

4.7.2.3 Hardness of Top Squeeze Roll

The bottom squeezing roller is made up of stainless steel. The top squeezing roller has a metallic core part which is covered with synthetic material. If the hardness of the top roller is low, then there will be more flattening of the roller. Thus the contact area between the rollers will increase and the pressure acting at the nip zone will decrease. Therefore, the size pick-up will increase. In contrast, harder top roller will produce sharper nip and lower wet pick-up. The shore hardness of the top roller is around 45°.

4.7.2.4 Thickness of Synthetic Rubber on the Top Roller

If the thickness of the synthetic rubber cover on the top roller is higher, then the extent of flattening will be more. This will reduce the nip pressure and thus the wet pick-up will increase.

4.7.2.5 Position of Immersion Roller

The position of the immersion roller within the size box is adjustable. If the height of immersion roller is lowered, then the residence time of the warp sheet within the size paste will increase. This will lead to increased wet pick-up if other factors remain constant.

4.7.2.6 Speed of Sizing

Speed of sizing also influences the wet pick-up by the warp sheet. This can be summarized as follows:

- Higher speed reduces the residence time of the warp sheet within the size paste. This should reduce the wet pick-up.

- Higher speed increases the drag force between the warp sheet and size paste, which should induce more flow of paste with the warp sheet. This should increase the wet pick-up.

- Higher speed reduces the time of squeezing. This should also increase the wet pick-up.

The aforesaid factors have contradictory influences on wet pick-up. Therefore, the speed of sizing will influence the wet pick-up based on the preponderance of aforesaid factors. Generally, wet pick-up increases with the increase in sizing machine speed. However, in modern sizing machines, the squeeze pressure is automatically adjusted with the change in machine speed so that the desired level of wet pick-up is maintained.

In modern sizing machines, the practical speed is around 100 m/min, though machine manufacturers claim that the speed can be as high as 150 m/min.

4.7.2.7 Percent Occupation and Equivalent Yarn Diameter

The relative closeness of the yarns inside the size box is expressed by percentage occupation and equivalent yarn diameter. A 100% occupation implies that the yarns are physically touching each other. This is not at all preferable as the size encapsulation will not occur at the place where two yarns are touching each other. The total number of yarns in the size box for 100% occupation can be calculated if the nominal yarn diameter is known. Equivalent yarn diameter indicates the space between the two yarns in terms of yarn diameter. If the equivalent yarn diameter is zero that means that the yarns are touching each other, that is 100% occupation. Figure 4.23 depicts the situation with 100% occupation, that is zero equivalent yarn diameter. Figure 4.24 presents a situation with 50% occupation, that is equivalent yarn diameter of one. If the equivalent yarn diameter is two, then percent occupation is 33%.

The expression for percent occupation is given as follows:

Percent occupation

$$= \frac{\text{Actual number of yarns in sizing machine} \times 100}{\text{Number of yarns in sizing machine with 100\% occupation}}$$

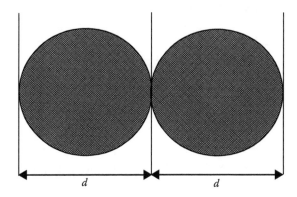

FIGURE 4.23 Yarn arrangement for 100% occupation.

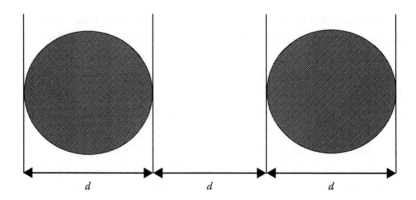

FIGURE 4.24 Yarn arrangement for 50% occupation.

The percent occupation and equivalent yarn diameter are related with the following expression:

$$\text{Percent occupation} = \frac{100}{1 + \text{Equivalent yarn diameter}} \qquad (4.3)$$

Generally a percent occupation of 50%–60% is considered acceptable. If the percent occupation is very high (>75%), then the yarns may not be uniformly coated by the size film. For a warp sheet having very large number of yarns, it may be preferable to use two size boxes to keep the percent occupation value within the permissible range.

4.7.2.8 Sizing Diagram

Figure 4.25 presents the schematic relationship between the concentration of size paste and size add-on %.

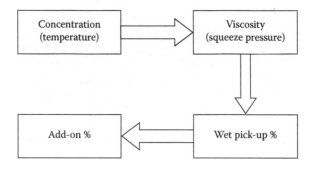

FIGURE 4.25 Relationship between concentration and add-on %.

- At a given temperature, the concentration of size paste determines the viscosity, that is *viscosity = f(concentration)*

- For a given yarn, squeeze pressure and sizing speed, the viscosity of size paste determines the wet pick-up, that is *wet pick-up = f(viscosity)*

- Wet pick-up determines the size add-on, that is *add-on = f(wet pick-up)*. For a given concentration of size paste, higher wet pick-up leads to higher add-on and vice versa.

- Therefore, if other conditions are same, then there will be a functional relationship between size paste concentration and size add-on. So *add-on = f(concentration)*.

Figure 4.26 depicts the relationship between concentration, viscosity, wet pick-up and add-on (Ormedrod and Sondhelm, 1995). Thin boiling starch requires higher level of concentration than the normal starch for the same level of viscosity. Therefore, even if the wet pick-up is same, add-on will be higher (due to higher concentration) for thin boiling starch. Besides, for the same level of add-on, wet pick-up will be less for thin boiling starch. Therefore, water evaporation during drying will be lower for thin boiling starch as compared to normal starch. This will lead to energy saving in the case of sizing with thin boiling starch. In Figure 4.26, the first, second, third and fourth quadrants indicate the relations between viscosity-wet pick-up, wet

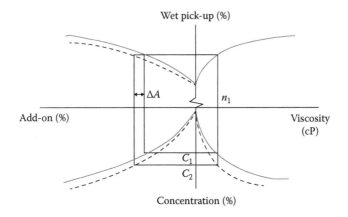

FIGURE 4.26   Sizing diagram.

pick-up-add-on, add-on-concentration and concentration-viscosity, respectively. It is seen in the fourth quadrant that thin boiling starch requires higher concentration (C_2) than normal starch (C_1) for creating the same level of size paste viscosity (η_1). This will lead to the same wet pick-up for both types of starch. This is indicated by only one curve in the first quadrant. Now, add-on will be more for the thin boiling starch (ΔA) even at the same level of wet pick-up. Thus, in the second quadrant, the broken line representing the thin boiling starch is positioned below the solid line representing normal starch. The third quadrant shows the intended relationships between size paste concentration and add-on.

4.7.2.9 Crowning of Top Roller

High-pressure squeezing is used to reduce the load on the drying system. In modern sizing machines, the squeezing force can go up to 100,000 N (100 kN). This force is applied on the two sides of the metallic core of top squeeze roller. This pressure is good enough to cause bending in the top squeeze roller, which may result in uneven pressure along the nip line. To overcome this problem, crowned top rollers are used. The synthetic rubber–coated top squeeze roller is manufactured in such a way that the diameter at the sides is lower as compared to that of at the middle as shown in Figure 4.27. This is compensated by the bending of the top rollers and uniform pressure is obtained along the nip line. Carbon paper impression of squeeze roller nip pressure is often taken in the industry to check the variation of pressure along the nip line.

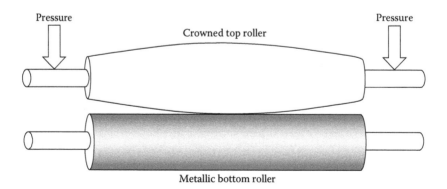

FIGURE 4.27   Crowning of a top roller.

4.7.3 Drying Zone

This is the zone where the wet warp sheet is dried by evaporating the water. The drying operation is very crucial because of the following reasons:

1. It consumes most of the energy of sizing process.

2. Inadequate drying will cause sticking of yarns with one another causing problems in weaving.

3. Over-drying will make the size film brittle and therefore it may shed easily during abrasion or bending.

Drying is done by passing the warp sheet over multiple drying cylinders coated with Teflon (polytetrafluoroethylene) and arranged in a sequential manner. The number of drying cylinders can vary from 2 to 30 depending on the amount of water to be evaporated in unit time. In general, higher speed of sizing would require more number of drying cylinders. The following expression is useful for calculating the mass of water to be evaporated during drying:

$$
\begin{aligned}
&\text{Mass (kg) of water to be evaporated} \\
&\text{per unit of oven dry mass (kg) of yarn} \\
&= \left(\frac{\text{Add-on \%}}{\text{Concentration \%}} \right) - \left(\frac{\text{Add-on \%}}{100} \right)
\end{aligned}
\tag{4.4}
$$

The first part of the expression yields wet pick-up. If the mass of dry size is subtracted from the wet pick-up, then the amount of water to be evaporated can be obtained. The aforementioned equation presumes that there is no residual moisture in the sized yarn after drying. The results obtained from the above expression is depicted in Figure 4.28. For the same level of size add-on, more water has to be evaporated if the size paste concentration is low. However, for a running machine, it is more important to calculate the mass of water to be evaporated in unit time (minute). This will be dependent on the following factors:

• Sizing machine speed

• Total number of yarns

• Linear density of yarns (tex) excluding moisture

• Add-on %

• Size paste concentration %

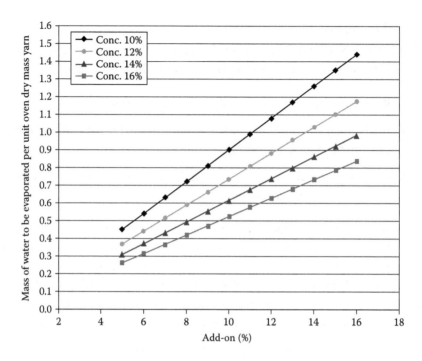

FIGURE 4.28 Effect of add-on % and concentration % on the amount of water evaporation.

The oven dry mass of yarns passing through the machine per minute can be expressed as follows.

$$= \frac{\text{Sizing machine speed (m/min)} \times \text{Total number of yarns} \times \text{tex}}{1000 \times 1000} \text{ kg}$$

The mass of paste picked up by the warp sheet per minute will be

$$= \left(\frac{\text{Sizing machine speed (m/min)} \times \text{Total number of yarns} \times \text{tex}}{1000 \times 1000} \right.$$

$$\left. \times \text{ Wet pick-up} \right) \text{kg}$$

$$= \left(\frac{\text{Sizing machine speed (m/min)} \times \text{Total number of yarns} \times \text{tex}}{1000 \times 1000} \right.$$

$$\left. \times \frac{\text{Add-on \%}}{\text{Concentration \%}} \right) \text{kg}$$

The mass of water to be evaporated per minute will be

$$= \frac{\text{Sizing machine speed (m/min)} \times \text{Total number of yarns} \times \text{tex}}{1000 \times 1000}$$

$$\times \frac{\text{Add-on \%}}{\text{Concentration \%}} \times \left(1 - \frac{\text{Concentration \%}}{100}\right) \text{kg} \qquad (4.5)$$

4.7.3.1 Methods of Drying

The methods of drying in sizing process can broadly be divided into the following categories:

- Conduction

- Convection

- Infrared

In conduction method, the warp sheet is passed over metallic cylinders, which are heated using superheated steam. Heat exchange takes place between the wet warp sheet and heated cylinders, and in the process, the warp sheet is dried (Figure 4.29). The energy efficiency of this process is very high. The problem with this system is that only one side of the warp sheet is exposed to the heated cylinder at a time. This problem can be overcome in convection method. In convection method, hot air is circulated within an enclosed chamber and the warp sheet is passed through the chamber with the help of some guides (Figure 4.30). Both the sides of the warp sheet are exposed to the hot air at the same time, which ensures uniform drying. However, the energy efficiency of the process is lower as compared to that of conduction process. Infrared drying can replace the steam-based drying system. Infrared dryers transfer heat by sending waves of electromagnetic energy to the sized yarns. The source of the energy is either electricity or gas. In the case of the former, a series of infrared lamps and reflectors are used to heat the yarns. The wavelength of the infrared light can be from 1.2 to 5.4 μm and it is inversely proportional with the temperature of the source. Infrared drying is yet to become popular in sizing.

To ensure better drying and reduction of load on individual cylinders, the wet warp sheet is often split into multiple sheets. Each sub-sheet is then dried by a separate group of drying cylinders (Figure 4.31). This initial

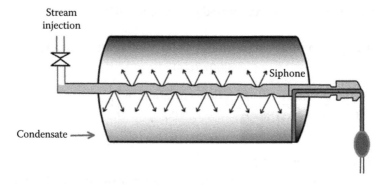

FIGURE 4.29   Conduction drying.

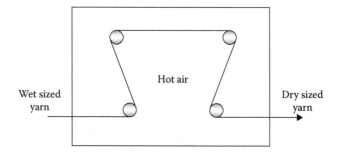

FIGURE 4.30   Convection drying.

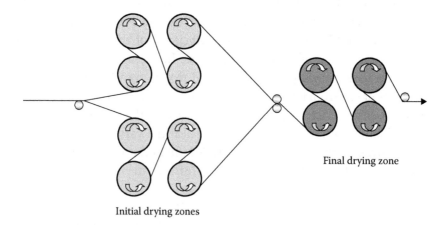

FIGURE 4.31   Two zone drying.

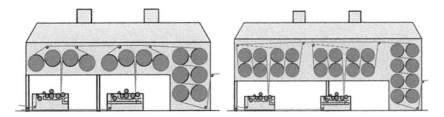

FIGURE 4.32 Drying zone configurations in a sizing machine. (Source: Technical literature of TTS20S spun sizing machine, T-Tech Japan Corp.)

drying is generally done at relatively lower temperature. Subsequently, all the sub-sheets are merged together and final drying takes place using another set of drying cylinders. The temperature range of drying cylinders for cotton warp is 100°C–140°C.

Various configurations of drying zone is possible to fulfil the diverse drying need for different types of warp sheet and end densities. Figure 4.32 depicts two possible drying zone configurations in a sizing machine.

4.7.3.2 Splitting

After drying, the warp sheet is split so that the yarns regain their individual identity before they are wound on the weaver's beam. This is depicted in Figure 4.33. Splitting is required because the warp yarns, coming out from the drying section, adhere to each other depending on the efficiency of the pre-drying section.

Lease rods, which are often coated with chromium, are used to split the warp sheet in a systematic manner as shown in Figures 4.34 and 4.35. Figure 4.34 depicts a situation when each of the lease rods split

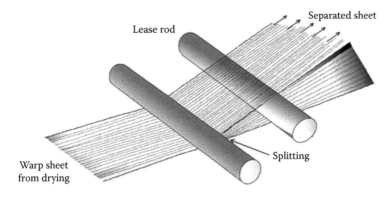

FIGURE 4.33 Splitting of a warp sheet.

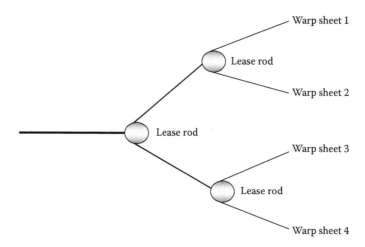

FIGURE 4.34   Splitting of a sized warp sheet into equal parts.

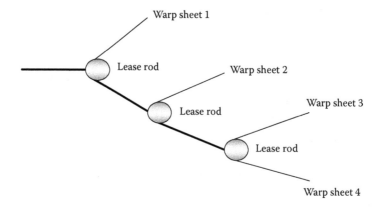

FIGURE 4.35   Splitting of a sized warp sheet into unequal parts.

the incoming warp sheet into two equal sub-sheets. On the other hand, Figure 4.35 presents a situation where warp sheet originating from a particular warper's beam is separated at a time by a lease rod. So, if it is assumed that warp sheet is originating from four warper's beams placed on the creel, the first lease rod will split the warp sheet into two unequal parts. One part will have yarns from a particular warper's beam, whereas the other part will consist of yarns originating from the remaining three warper's beam. The number of lease rods is always equal to $(n-1)$, where n is the number in which the warp sheet is divided after splitting.

The function of the lease rods is to separate the individual yarns which are sticking together by dried size film. During the splitting,

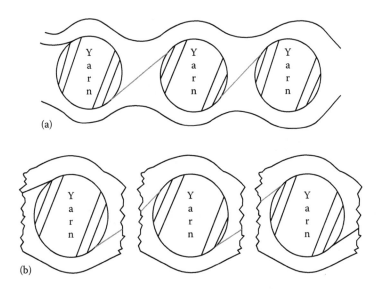

(a)

(b)

FIGURE 4.36 Arrangement of hairs in yarns (a) before and (b) after splitting.

some amount of size film drops as waste. However, a large number of longer protruding fibres, bridging two adjacent yarns, also break into smaller pieces. Therefore, splitting is considered to have some beneficial effect from hairiness viewpoint. This is represented pictorially in Figure 4.36.

In some of the sizing machines, splitting is done before drying zone. This technique is known as wet splitting. In this case the size materials remain in wet condition when the splitting is done. Therefore, the force required to split the warp is much lower as compared to that of dry splitting.

4.7.4 Headstock Zone

After the splitting, the warp sheet is finally wound on the weaver's beam. The warp sheet passes through an adjustable reed which can be expanded or collapsed based on the end density required on the weaver's beam.

4.8 PRE-WETTING OF YARNS

Pre-wetting is often done for the staple spun yarns to make the sizing process more efficient. The warp sheet is passed through a box which contains hot water (temperature around 90°C) and thus the waxes and other impurities are partially removed from the yarns (Figure 4.37). This improves the adhesion between the yarns and the size film. If the sizing machine

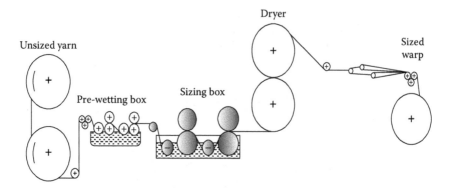

FIGURE 4.37   Simplified representation of pre-wetting process.

has two size boxes, then the first one can be used for pre-wetting and the second one for sizing. Generally, size boxes having two dip and two nip are preferred when pre-wetting is done. After pre-wetting, water occupies the core of the yarns and thus the penetration of size within the yarn structure reduces and uniform size film is formed over the yarn surface. High squeezing pressure is used at the nip of pre-wetting box so that the water retained by the yarns is minimized. This precludes the possibility of dilution of paste concentration in the size box as well as reduction of paste temperature. The advantages of pre-wetting are as follows:

- Reduction of size ingredient consumption up to 50%

- Increase in yarn strength

- Reduction in yarn hairiness

- Improvement in loom efficiency

4.9 QUALITY EVALUATION OF SIZED YARNS

The real performance of sized yarn can be appraised only during the weaving operation. However, the performance of sized yarns can be forecasted by judging the following quality parameters:

- Tenacity and breaking elongation of sized yarn

- Cohesiveness and adhesion of sized film

- Abrasion resistance

- Fatigue resistance

TABLE 4.8 Properties of Starch, CMC and PVA

Mechanical Properties	Size Type		
	Starch	CMC	PVA
Tenacity (cN/mm²)	3.5	3.3	4.0
Elongation (%)	20.6	30.3	45.2
Initial modulus (cN/mm²)	200.9	150.2	90.8

Source: Maatoug, S. et al., *AUTEX Res. J.*, 8, 239, 2007.

Tensile strength (tenacity) enhancement after sizing does not exhibit good correlation with the actual warp breakage rate during weaving, as the tension acting on yarn during weaving does not exceed 20% of the yarn breaking strength. Cohesiveness of size film is evaluated by measuring the tensile properties of a thin film made from the size paste. A comparison of cohesive properties of maize starch, PVA and carboxymethyl cellulose (CMC) films is given in Table 4.8.

PVA demonstrates higher cohesiveness and elongation than maize starch and CMC. However, PVA is expensive than starch. Therefore, a mixture of starch and PVA is often used for the sizing of spun yarns as it gives optimized weaving performance and sizing cost.

Adhesive power indicates the compatibility of the size material with the fibre. It can be measured using the following expression:

$$\text{Adhesive power} = \frac{\text{Breaking strength of sized roving at gauge length } l}{\text{Breaking strength of sized roving at zero gauge length}}$$

l should be greater than staple length of fibre.

If the adhesion between the fibres and size film is good, then the slippage of the fibres in the sized roving during tensile testing will reduce. This will increase the adhesive power as the breaking strength of sized roving at gauge length l will increase. When adhesion is good, the critical adhesive power is attained at lower add-on % as indicated by the solid line in Figure 4.38. In contrast, if the adhesive power is poor, then higher add-on % is required to reach the level of critical adhesive power as indicated by the dotted line in Figure 4.38.

During weaving, abrasion takes place between warp yarns and loom parts. Sized yarns can be subjected to abrasion tests and the number of cycles required to break a given number of yarns is noted. These observations can be used to calculate the mean abrasion cycle that can be resisted

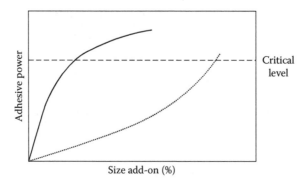

FIGURE 4.38 Size add-on % vs adhesive power.

by a sized yarn. Alternatively, sized yarns can be subjected to a fixed number of abrasion cycles and then the % deterioration in tensile strength can be calculated using the following expression. Lower deterioration implies good quality of sizing and vice versa.

$$\begin{array}{l} \text{\% Deterioration} \\ \text{due to abrasion} \end{array} = \dfrac{\begin{array}{c}\text{Original breaking strength of yarn} \\ -\text{Breaking strength of yarn after abrasion}\end{array}}{\text{Original breaking strength of yarn}} \times 100$$

During weaving, the warp yarns undergo repeated extension and bending. This causes cumulative damage to the fibres and yarn structure. As a result the yarn actually fails, due to fatigue, at a breaking load, which is much lower than its actual breaking load. Even a very strong metal wire breaks after repeated flexing due to poor fatigue resistance. Sulzer-Ruti Webtester is used to evaluate the fatigue resistance of sized yarns. The instrument simulates most of the stresses that act upon warp yarns during the weaving.

NUMERICAL PROBLEMS

4.1 One hundred kg of oven dry warp yarns were sized to the add-on of 8% and dried to overall (yarn and size film) moisture content of 10%. Calculate the final mass of the sized yarns.

Solution:
Mass of oven dry warp = 100 kg
Add-on = 8%

So

$$\frac{\text{Oven dry mass of size materials}}{\text{Oven dry mass of warp}} \times 100 = 8$$

or

$$\frac{\text{Oven dry mass of size materials}}{100} \times 100 = 8$$

or

$$\text{Oven dry mass of size} = 8 \text{ kg}$$

Total oven dry mass of materials (size + warp yarns) = 108 kg

$$\text{Moisture content} = 10\%$$

So

$$10 = \frac{W}{W + 108} \times 100$$

where W is mass of water or, W = 12 kg

Oven dry mass of yarn + oven dry size + overall moisture = (100 + 8 + 12) kg

So final mass of sized yarn is 120 kg

4.2 A lot of 20 kg cotton warp, having moisture regain of 8.5%, is sized with a paste of 15% concentration. If 10% add-on is aimed, then what should be the wet pick-up? How much water has to be evaporated so as to leave 8% moisture content in warp and size film?

Solution:
Moisture regain (MR) in cotton warp is 8.5%.
So

$$\frac{W}{D} \times 100 = 8.5$$

where
W is the mass of water
D is the oven dry mass of warp yarns

or

$$W = 0.085 \times D$$

Total mass of input warp = Oven dry mass of warp

+ Mass of water in warp

or

$$D + 0.085D = 20 \text{ kg}$$

or

$$D = 18.43 \text{ kg}$$

$$\text{Oven dry mass of size added} = D \times \frac{\text{Add-on \%}}{100} = 18.43 \times \frac{10}{100} = 1.843 \text{ kg}$$

Now,

Mass of water (W) in the input warp $= (20 - 18.43) \text{ kg} = 1.57 \text{ kg}$

$$\text{Wet pick-up (WPU)} = \frac{\text{Add on \%}}{\text{Concentration \%}} = 10/15 = 0.667$$

Total mass of size paste picked up by warp $= \text{WPU} \times D = 0.667 \times 18.43 \text{ kg}$

$$= 12.29 \text{ kg}$$

Mass of water within the size paste picked up by warp

$$= 12.29 \times \left(\frac{100 - \text{Concentration \%}}{100} \right)$$

$$= 12.29 \times \left(\frac{100 - 15}{100} \right)$$

$$= 10.45 \text{ kg}$$

Total mass of water in warp after size paste pick-up

= Mass of water in size paste picked up by warp

+ Mass of water originally present in input warp

$$= (10.45 + 1.57)\ \text{kg}$$

$$= 12.02\ \text{kg}$$

Water to be retained (for 8% moisture content in yarn and size film)

$$= \frac{8}{92} \times 18.43 + \frac{1.843 \times 8}{92}$$

$$= 1.76\ \text{kg}$$

Therefore,

The mass of water to be evaporated $= (12.02 - 1.76)\ \text{kg} = 10.26\ \text{kg}$

So 10.26 kg of water has to be evaporated.

4.3 A 40 tex cotton yarn has add-on of 8%. If the moisture regain of the input warp is 10%, then determine the oven dry mass of the size added per kg of unsized warp.

Solution:
Size add-on = 8%, Moisture regain = 10%
So

$$\frac{W}{D} \times 100 = 10$$

where
   W is the mass of water
   D is the oven dry mass of yarn

or

$$W = 0.1D$$

Let the total mass of unsized yarn = 1 kg

So

$$D + W = 1 \text{ kg, or, } D + 0.1D = 1 \text{ kg}$$

or

$$D = 0.909 \text{ kg}$$

Now, add-on is 8%.
So

$$\frac{\text{Oven dry mass of size materials}}{\text{Oven dry mass of yarn } (D)} \times 100 = 8$$

or

Oven dry mass of size materials $= D \times 0.08 = 0.909 \times 0.08 = 0.073 \text{ kg} = 73 \text{ g}$

So oven dry mass of size materials per kg of unsized warp is 73 g.

4.4 The stretch in a sizing machine processing acrylic warp yarn is 1%, 3% and 2% in the creel zone, sizing zone and drying zone, respectively. If the warp crimp in the woven fabric is 10%, then determine the length of fabric that can be produced from 10,000 m length of supply warp sheet.

Solution:
Total stretch in sizing can be calculated using the following multiplicative expression:

$$\text{Total stretch} = \left[\left(1 + \frac{S_1}{100}\right) \times \left(1 + \frac{S_2}{100}\right) \times \left(1 + \frac{S_3}{100}\right) - 1 \right] \quad (4.6)$$

where S_1, S_2 and S_3 are the % stretches in the three zones
So

$$\text{Total stretch} = \left[\left(1 + \frac{1}{100}\right) \times \left(1 + \frac{3}{100}\right) \times \left(1 + \frac{2}{100}\right) - 1 \right]$$

$$= 0.061 = 6.1\%$$

So

$$\text{Length of sized yarn} = (L_y) = 1.061 \times 10,000 = 10,610 \text{ m}$$

$$\text{Crimp \%} = 10\%$$

So

$$10 = \frac{\text{Length of yarn } (L_y) - \text{Length of fabric } (L_f)}{\text{Length of fabric } (L_f)} \times 100$$

Or

$$L_f = \frac{L_y}{1.1} = \frac{10,610 \text{ m}}{1.1} = 9,645.5 \text{ m}$$

So the length of woven fabric will be 9645.5 m.

4.5 A sizing machine is running at 100 m/min speed with 6000 ends. The add-on requirement is 12% and the concentration of size paste is 18%. If the yarn count is 20 tex (without any moisture) and residual moisture content in the sized yarn and film after drying is 10%, then calculate the number of drying cylinders required. One drying cylinder can evaporate 3 kg water per minute.

Solution:

Oven dry mass of warp passing through the machine/min

$$= \frac{100 \times 6000 \times 20}{1000 \times 1000} \text{ kg} = 12 \text{ kg}$$

$$\text{Wet pick-up } (WPU) = \frac{\text{Add-on \%}}{\text{Concentration \%}} = \frac{12}{18} = 0.667$$

Total mass of size paste picked up by the warp

$$= WPU \times \text{oven dry mass of warp}$$

$$= \frac{12}{18} \times 12 \text{ kg} = 8 \text{ kg}$$

Oven dry mass of size = Mass of size paste × Concentration = 8×0.18

Total oven dry mass of warp and size = $(12 + 8 \times 0.18)$ kg = 13.44 kg

For moisture content of 10%, mass of water to be retained in warp

and size film $= \frac{1}{9} \times 13.44$ kg

$$= 1.49 \text{ kg}$$

Mass of water in picked up size paste of 8 kg

$$= 8 \times \left(\frac{100 - \text{Concentration } \%}{100} \right) = 6.56 \text{ kg}$$

So water to be evaporated per minute = (6.56 – 1.49) kg = 5.07 kg

Drying capacity of one cylinder = 3 kg/min

Number of cylinders will be = 5.07/3 ≈ 2

So the required number of drying cylinders is 2.

4.6 Determine the failure rate of sized yarns if the six yarns tested for abrasion breaks ruptured after 1025, 1550, 2232, 3785, 5608 and 7918 cycles. What is the reliability of the yarns at 2000 cycles if the reliability curve is exponential?

Solution:
Reliability is defined as the probability of survival of a product after a certain time. At time zero, the reliability of any product is one as it has 100% probability for survival. On the other hand, when time approaches infinity, the reliability tends to be zero as the probability of survival of any product becomes zero. So reliability is a function of time.

Failure rate (ω) = Number of failure/cycle

$$= \frac{\text{Total number of failures}}{\text{Total number of cycles}}$$

$$= \frac{6}{1025 + 1550 + 2232 + 3785 + 5608 + 7918}$$

$$= 2.712 \times 10^{-4} \text{ per cycle}$$

Average life is the reciprocal of failure rate.

So, average life (θ) $= \frac{1}{\omega} = 3686$ cycle

According to exponential reliability, the reliability decays exponentially with the increase in time (t).

Here, probability density function (PDF) of failure is $= \dfrac{1}{\theta} e^{-\frac{t}{\theta}}$

Cumulative density function or CDF(failure)

$$= \int_0^t \frac{1}{\theta} e^{-\frac{t}{\theta}} = \left(-e^{-\frac{t}{\theta}} \right)_0^t = \left(-e^{-\frac{t}{\theta}} + 1 \right)$$

so

Cumulative density function CDF (survival)

$$= 1 - \text{CDF(failure)} = 1 - \left(-e^{-\frac{t}{\theta}} + 1 \right) = e^{-\frac{t}{\theta}} \tag{4.7}$$

so

Reliability $= e^{-\frac{t}{\theta}}$ (where t is the cycle number in this case)

$$= e^{-\frac{2000}{3686}}$$

$$= 0.581$$

So reliability of the yarn at 2000 cycle is 0.581.

REFERENCES

Banerjee, P. K. 2015. *Principles of Fabric Formation*. Boca Raton, FL: CRC Press.

Goswami, B. C., Anandjiwala, R. and Hall, D. M. 2004. *Textile Sizing*. New York: Marcel Dekker, Inc.

Hari, P. K., Behera, B. K., Prakash, J. and Dhawan, K. 1989. High pressure squeezing in sizing: Performance of cotton yarn. *Textile Research Journal*, 59: 597–600.

Maatoug, S., Ladhari, N. and Sakli, F. 2007. Evaluation of weaveability of sized cotton warps. *AUTEX Research Journal*, 8: 239–244.

Ormerod, A. and Sondhelm, W. S. 1995. *Weaving: Technology and Operations*. Manchester, UK: The Textiles Institute.

Technical literature of Selvol™ Polyvinyl Alcohol for Textile Warp Sizing, 2011, Sekisui Specialty Chemicals, Osaka, Japan, www.sekisui-sc.com. Accessed on 21st March, 2016.

Technical literature of TTS20S spun sizing machine, T-Tech Japan Corp., Ishikawa, Japan, www.t-techjapan.co.jp. Accessed on 21st March, 2016.

CHAPTER **5**

Weave Design

5.1 INTRODUCTION TO WEAVE DESIGN

Fabric weave design implies the pattern of interlacement between the warp and weft yarns. The design influences aesthetics as well as properties of woven fabrics. Weave design is constructed on point paper using crosses (×) and blanks. A cross means that the warp yarn or end is moving over the weft yarn or pick. A blank means that the end is moving below the pick. In point paper design, ends are shown along the columns, whereas picks are shown along the rows. If a weave design repeats on 10 ends and 10 picks, then a 10×10 area on point paper is sufficient to represent the weave repeat.

The design of woven fabrics can be manipulated by changing the following two things:

1. Drafting

2. Lifting plan

5.1.1 Drafting

Drafting determines the allocation of ends to healds, that is which end will be controlled by which heald. Generally, drafting is made in such a way that the minimum number of healds is required to produce a particular design. This implies that ends having similar interlacement pattern should be controlled by the same heald. In the case of drafting, ends and healds are indicated along the columns and rows, respectively. A cross means that the end in column has been assigned to the heald in row. As per convention, drafting is shown above the weave design.

5.1.2 Lifting Plan

Lifting plan shows the position of healds (up or down) for different peaks, that is which heald or healds to be lifted in which pick. Lifting plan is dependent on the design and the drafting. In the case of lifting plan, healds and picks are indicated along columns and rows, respectively. A cross means that the heald in column is up (lifted or raised) when the pick in row is inserted. As per convention, lifting plan is shown on the right-hand side of weave design.

5.2 TYPES OF DRAFT

The following drafts are very commonly used for weaving the basic designs:

- Straight draft

- Pointed draft

- Skip draft

5.2.1 Straight Draft

In the case of straight draft, a diagonal line is created by the crosses (Figure 5.1). Generally, this implies that end 1 is controlled by heald 1, end 2 is controlled by heald 2 and so on.

5.2.2 Pointed Draft

In the case of pointed draft, a pointed line is created by the crosses (Figure 5.2). The repeat of the design contains more than one ends with similar interlacement pattern. For example, in Figure 5.2, the interlacement pattern is the same for ends 1 and 7, and thus, they are allocated to one heald (heald number 1). It is also true for ends 2 and 6, 3 and 5, and 4 and 8.

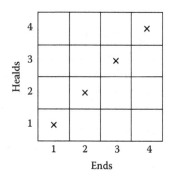

FIGURE 5.1 Straight draft.

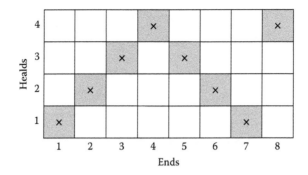

FIGURE 5.2   Pointed draft.

Therefore, this design, which has eight ends in the repeat, requires only four healds. Pointed twill weaves are made using pointed draft.

5.2.3 Skip Draft

In the case of skip draft, two or more healds are controlled by a single shedding cam. Plain woven fabrics can be produced with two healds. However, for heavy (high areal density) plain woven fabrics, the number of ends is very high. It often becomes convenient to use four healds for the heavy plain woven fabrics. Therefore, the number of ends controlled by a single heald becomes less as compared to the situation with only two healds. The skip draft for plain woven fabrics with four healds is shown in Figure 5.3.

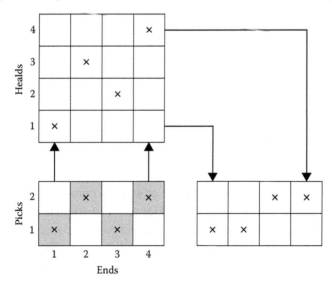

FIGURE 5.3   Skip draft.

TABLE 5.1   Allocation of Healds for Skip Draft

End Number	Heald Number
1	1
2	3
3	2
4	4

The allocation of healds is shown in Table 5.1.

So the allocation of healds for ends 2 and 3 differs from that of straight draft. Lifting plan shows that for the 1st pick, healds 1 and 2 are in up position and healds 3 and 4 are in down position. For pick 2, healds 1 and 2 are in down position and healds 3 and 4 are in up position. So the movement pattern of healds 1 and 2 is same and the movement pattern of healds 3 and 4 is also same. Besides, the movement pattern of healds 3 and 4 is just opposite as compared to that of healds 1 and 2. Therefore, healds 1 and 2 can be tied or coupled together with ropes or strings, and their movements during shedding operation can be controlled by a single cam. Similarly, movements during shedding operation of healds 3 and 4 can be controlled by another cam. This becomes possible as healds 1 and 2 are physically close to each other and healds 3 and 4 are also physically close to each other. Thus, skip drafting helps reduce the number of mechanical components (cam, follower, treadle lever, etc.) in the loom.

5.3 BASIC WEAVES

The following weaves are the three basic weaves. All other weaves are derivatives of these three.

1. Plain weave and its derivatives

2. Twill weave

3. Satin and sateen weave

5.3.1 Plain Weave

Plain weave is the simplest possible and most commonly used weave. The repeat size is 2×2 as depicted in Figure 5.4. It implies that the weave repeats on two ends and two picks. It gives the maximum number of interlacement in the fabric and therefore the fabric becomes very firm. As the yarns have maximum possible interlacements, the crimp in the yarns is also higher as compared to other weaves. Figure 5.5 depicts the interlacement pattern in a plain woven fabric.

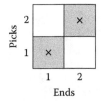

FIGURE 5.4　Point paper representation of a plain weave.

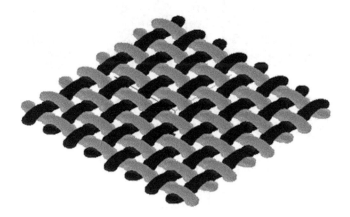

FIGURE 5.5　Interlacement pattern of a plain weave (warp: grey and weft: black).

Warp rib, weft rib and matt (basket) weaves are the derivatives of plain weave. All these designs can be woven with two healds.

5.3.1.1 Warp Rib
In the case of warp rib, two neighbouring picks move in a group as shown in Figure 5.6. Prominent ribs become visible in the warp direction of fabrics, which is created by the floats of the ends. The picks undergo more number of interlacement than the ends and therefore the crimp in the weft yarns is higher than that of warp yarns.

Due to the interlacement pattern, warp rib fabrics will have more tearing strength in the warp direction as compared to the plain woven fabrics having same yarns and thread density (ends per cm and picks per cm). For warp rib fabrics, two neighbouring picks will resist the tearing force together in a pair, resulting in higher tearing strength in warp direction as compared to equivalent plain woven fabrics.

The design, drafting and lifting plans of warp rib are shown in Figure 5.7. The design and lifting plan resemble each other.

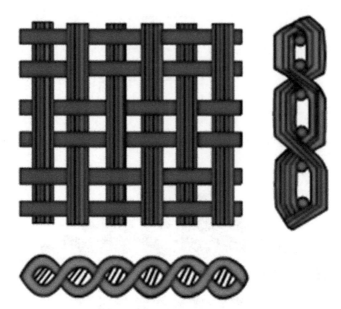

FIGURE 5.6   Interlacement pattern of a warp rib.

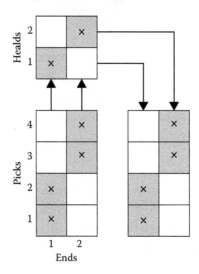

FIGURE 5.7   Design, drafting and lifting plan of a warp rib.

5.3.1.2 Weft Rib

In the case of weft rib, two neighbouring ends move in a group as shown in Figure 5.8. Prominent ribs become visible in the weft direction of fabric and they are created by the floats of the picks. The ends undergo more number of interlacements than the picks and therefore the crimp in the warp yarns is

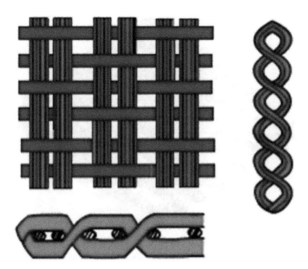

FIGURE 5.8   Interlacement pattern of a weft rib.

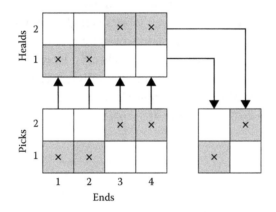

FIGURE 5.9   Design, drafting and lifting plan of a weft rib.

higher than that of weft yarns. Weft rib fabrics will have more tearing strength in the weft direction as compared to the equivalent plain woven fabrics.

The design, drafting and lifting plans of weft rib are shown in Figure 5.9. The design and drafting resemble each other.

5.3.1.3 Matt or Basket Weave

In a matt weave, multiple ends and picks interlace with each other in groups following the pattern of plain weave. The number of interlacements in the fabric is much lower than that of plain weave. In a 2×2 matt weave, two neighbouring ends and two neighbouring picks form pairs and

FIGURE 5.10   Interlacement pattern of a 2×2 matt weave (warp: grey and weft: black).

interlace in the form of plain weave as shown in Figure 5.10. Therefore, the tearing strength of matt woven fabrics is higher in both directions as compared to that of equivalent plain woven fabrics. It is possible to weave 3×3 and 4×4 matt weaves using only two healds. However, as the number of yarns in a group increases, the number of interlacements in the fabric reduces. Therefore, the fabric becomes less firm. The design, drafting and lifting plans of a 2×2 matt weave are shown in Figure 5.11.

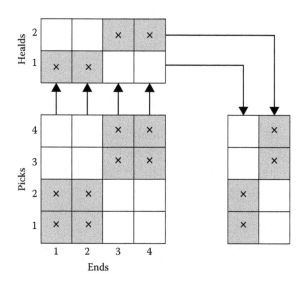

FIGURE 5.11   Design, drafting and lifting plan of a 2×2 matt weave.

5.3.2 Twill Weave

Twill weaves are characterised by a diagonal line in the fabric, which is created by floats of ends or picks (Grosicki, 1997). The simplest twill weave is two up one down, that is 2/1 (or one up two down, i.e. 1/2), which repeats on three ends and three picks. Based on the prominence of warp or weft floats, twill weaves are classified as follows:

1. Warp-faced: 2/1, 3/1, 3/ 2, etc.

2. Weft-faced: 1/2, 1/3, 2/3, etc.

3. Balanced twill: 2/2, 3/3, 2/1/1/2, etc.

In a warp-faced twill, the floats of ends predominate over those of picks. In contrast, the floats of picks predominate over those of ends in weft-faced twill. In the case of balanced twill, the floats of ends and picks are equal. Figure 5.12 shows point paper design of a warp-faced (2/1) and a balanced (2/2) twill. In the case of a 2/1 twill, ends are in up position in six out of nine crossover points, implying that it is a warp-faced twill. On the other hand, in the case of a 2/2 twill weave, ends are in up position in eight out of sixteen crossover points, implying that it is a balanced twill. Figures 5.13 and 5.14 depict the interlacement pattern for 2/1 and 3/1 twill weaves, respectively. It can be seen from Figure 5.14 that there are long floats of warp (grey colour) over three consecutive picks visible at the face side of the fabric.

Twill weaves have lesser number of interlacements than the plain weave. Thus the crimp in yarns for twill weaves will be lower than that of plain weaves. For equivalent fabrics, 3/1 twill will give higher tearing strength, followed by 2/1 twill and plain weaves.

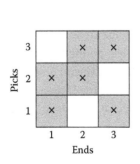
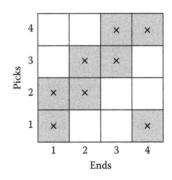

FIGURE 5.12 Warp faced (2/1) twill weave (left) and balanced (2/2) twill weave (right).

FIGURE 5.13   Interlacement pattern in a 2/1 twill weave (warp: grey and weft: black).

FIGURE 5.14   Interlacement pattern in a 3/1 twill weave (warp: grey and weft: black).

5.3.2.1 Pointed Twill

In pointed twills, there is no continuous twill line. However, the twill lines change directions at specified intervals and thus create pointed effect on the fabric. The design, drafting and lifting plan of a pointed twill based on basic 2/2 twill weave are shown in Figure 5.15. The fourth end is considered as the mirror line and the design is reversed such that the interlacement patterns of ends 5, 6 and 7 becomes identical with those of ends 3, 2 and 1, respectively. The interlacement patterns of end 4 and end 8 is same. The pointed twill is woven using the pointed draft as shown in Figure 5.15. The lifting plan resembles the left-hand side of the design, which is true for the pointed draft. Figure 5.16 depicts the extended view of the same pointed twill.

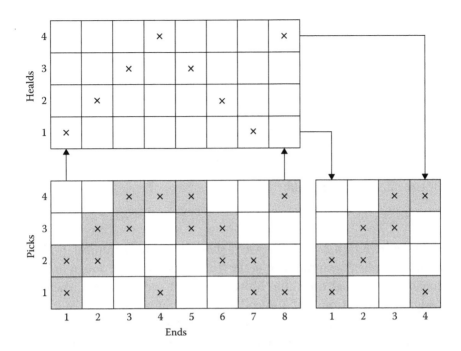

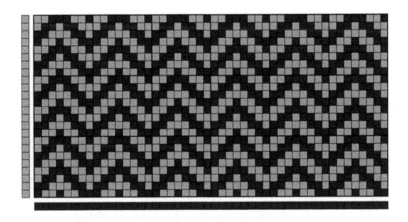

FIGURE 5.15   Design, drafting and lifting plan of a pointed twill.

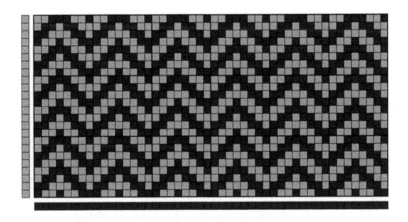

FIGURE 5.16   Extended view of a pointed twill.

5.3.2.2 Angle of Twill

The angle made by the twill line with the horizontal direction (weft direction) is known as angle of twill or twill angle (Figure 5.17). From point paper design, it seems that the angle will always be 45°. However, it is dependent on pick spacing, end spacing and move number of the design.

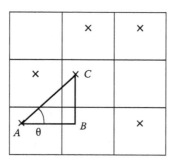

FIGURE 5.17   Angle of a twill.

In Figure 5.17, a 2/1 twill weave is shown with move number 1. Move number implies the movement of the starting point of the design in horizontal and vertical direction. Move number 1 indicates that when the design is shifted from end 1 to end 2, the starting point is moved by one step right and one step up, that is from A to C. Generally, for the construction of standard designs, move number 1 is used for both the directions. Here the angle θ (CAB) is the twill angle.

Therefore,

$$\theta = \tan^{-1}\left(\frac{BC}{AB}\right) = \tan^{-1}\left(\frac{\text{Pick spacing}}{\text{End spacing}}\right) = \tan^{-1}\left(\frac{p_2}{p_1}\right) \tag{5.1}$$

Therefore, the angle of twill depends on the ratio of pick spacing and end spacing as shown earlier. However, using move number 2 in the vertical direction, a steep twill can be produced where the angle of twill >45°. On the other hand, using higher move number in the horizontal direction, a reclined twill can be produced where the angle of twill <45°.

Thus the generalised expression for twill angle is as follows:

$$\text{Twill angle }(\theta) = \tan^{-1}\left(\frac{p_2}{p_1} \cdot \frac{\text{Move no. in vertical direction}}{\text{Move no. in horizontal direction}}\right) \tag{5.2}$$

Steep twill and reclined twill based on 3/1 twill weave are shown in Figure 5.18. Different twill angles are depicted in Figure 5.19.

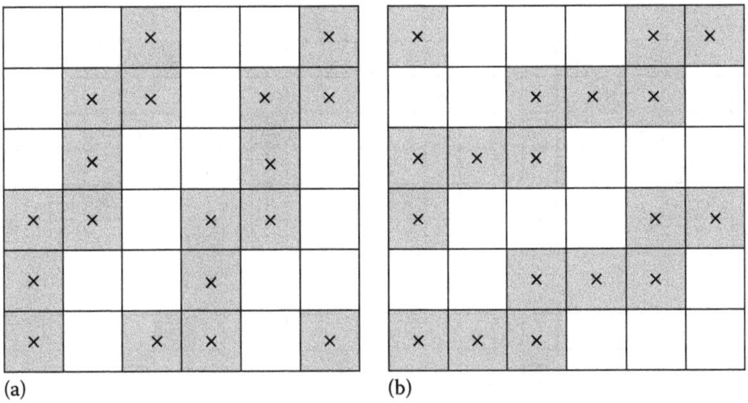

FIGURE 5.18 Steep and reclined twills. (a) Repeat size 3×6. (b) Repeat size 6×3.

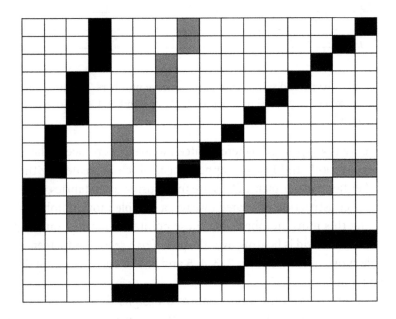

FIGURE 5.19 Different twill angles.

5.3.3 Satin and Sateen Weaves

Satin and sateen weaves are characterised by the following features:

- Only one crossover point in each end and pick within the repeat

- No continuous twill line

- Smooth appearance

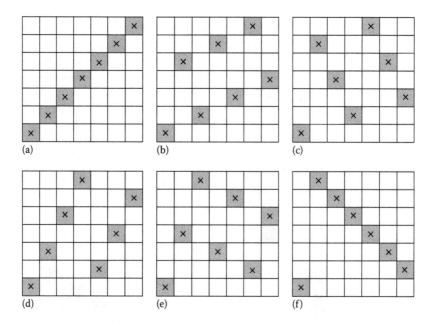

FIGURE 5.20 Seven-end sateen with various move numbers. (a) Move number 1. (b) Move number 2. (c) Move number 3. (d) Move number 4. (e) Move number 5. (f) Move number 6.

Satin weave is warp-faced, whereas sateen weave is weft-faced. The fabrics have very smooth and lustrous appearance, which is created by the long floats of either ends or picks. Flags are made using satin fabrics.

For the construction of a sateen weave, a feasible move number is chosen. Using this move number, only those points are marked on the point paper where the end is floating over the pick. For a seven-end sateen weave, probable move numbers are 1, 2, 3, 4, 5 and 6. The corresponding designs are shown in Figure 5.20.

It is observed from the aforementioned designs that move numbers 1 and 6 ($n - 1$, where n is the repeat size of the weave) produce twill weaves. However, move numbers 2, 3, 4 and 5 produce valid sateen weaves. On a point paper, sateen weave (weft-faced) can be converted to satin weave (warp-faced) by interchanging the crosses with blanks and vice versa. The interlacement pattern of five-end sateen, which has the smallest repeat size for a regular sateen, is depicted in Figure 5.21.

5.3.3.1 Six-End Regular Sateen

It has been demonstrated earlier that if move number is 1 or $n - 1$, then a twill weave is produced. Here, n is the repeat size of the design. If a six-end

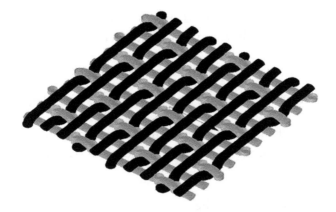

FIGURE 5.21 Five-end sateen (black weft is at the face side).

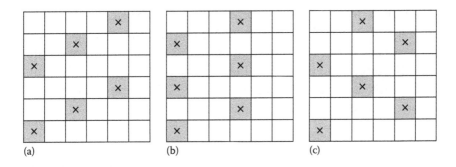

FIGURE 5.22 Six-end sateen. (a) Move number 2. (b) Move number 3. (c) Move number 4.

sateen weave is designed with move numbers 2, 3 or 4, then the following interlacement patterns will be produced (Figure 5.22).

In all the three cases, there are certain ends without any interlacement. Therefore, these designs are practically not valid. Therefore, six-end regular sateen (or satin) weave is not feasible.

5.3.3.2 Rules for Making Satin or Sateen Weaves

1. Move number 1 or $(n-1)$ should not be chosen as it will produce twill weaves (Robinson and Marks, 1967).

2. Move number and repeat size of the design should not have any common factor.

It seems from the point paper design that a satin will become sateen if the fabric is reversed (turned upside down). However, practically it is not true because satin fabric is warp-faced, and to make the effect of warp floats more prominent, the following steps are adopted:

- Use of coarser warp yarns than the weft yarns

- Use of higher ends per inch (*epi*) than the picks per inch (*ppi*) in fabric

Therefore, even if the fabric is reversed, the effect of weft floats will not be very prominent as the weft yarns are finer than warp yarns and *ppi* is lower than *epi*.

5.4 SOME FANCY WEAVES

5.4.1 Honeycomb

Honeycomb weaves show prominent diamond shapes on the fabrics created by the long floats of ends and picks. A honeycomb weave having a repeat size of 8×8 is shown in Figure 5.23 with drafting and lifting plan. The design can be produced with pointed draft and thus the lifting plan resembles the left-hand side of the design. The extended view of the Honeycomb weave is shown in Figure 5.24.

5.4.2 Mock Leno

In Mock leno weave, some of the ends have frequent interlacement, whereas the other ends have long floats. The fabric shows small holes created by the grouping of threads. A mock leno weave having a repeat size of 10×10 is shown in Figure 5.25 along with drafting and lifting plan. Only four healds are needed as the interlacement patterns of ends 1, 3, 5 are same and they are allocated to heald 1. Similarly, the interlacement patterns of ends 2 and 4 are same and they are assigned to heald 2 and so on. Figure 5.26 depicts the extended view of the Mock leno weave.

5.4.3 Huck-a-Back

Huck-a-back design has some similarity with Mock leno. A 10×10 Huck-a-back design is shown in Figure 5.27. If the design is divided in to four quadrants, then the top-right and bottom-left corners show similar interlacement pattern like Mock leno. However, the remaining two quadrants show plain weave–like interlacement pattern. Therefore, some of the ends (end number 2, 4, 7 and 9) will have long floats followed by regular interlacements. The design shown in Figure 5.27 requires four heald shafts. Figure 5.28 depicts the extended view of the Huck-a-back weave.

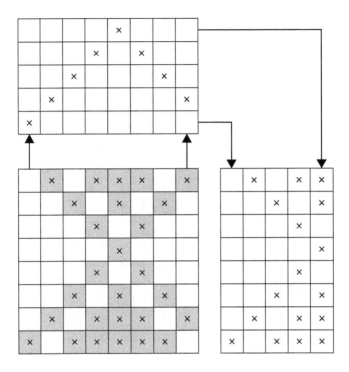

FIGURE 5.23 Design, drafting and lifting plan of a Honeycomb weave.

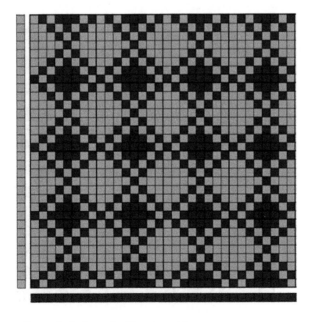

FIGURE 5.24 Extended view of a Honeycomb weave.

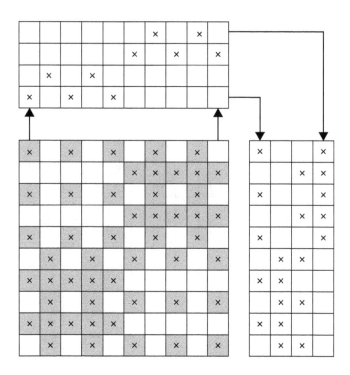

FIGURE 5.25 Design, drafting and lifting plan of a Mock leno weave.

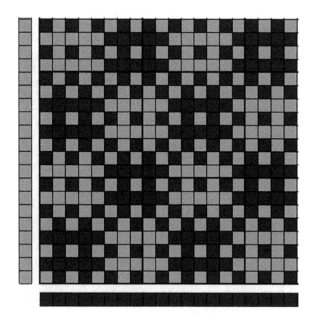

FIGURE 5.26 Extended view of a Mock leno weave.

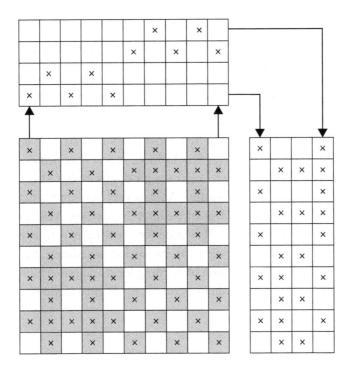

FIGURE 5.27 Design, drafting and lifting plan of a Huck-a-back weave.

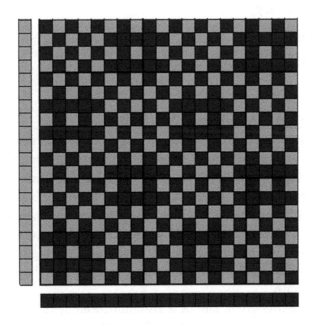

FIGURE 5.28 Extended view of a Huck-a-back weave.

5.5 COMPUTER-AIDED DESIGN

Computer-aided design (CAD) facilitates easy creation of any kind of weave and design and helps to simulate the created weave or design. The simulation is possible in the form of a virtual fabric in different colour combinations even before the actual fabric is produced. Textile CAD has almost unlimited features that can never be created by manual designing. This is now being adopted by almost every industry. To compete in the global market in an era of fast fashion, it has now become imperative to adopt CAD not only to increase productivity but also to achieve value addition in products. Textile CAD has very vast application in areas such as dobby, jacquard and print designs. In textile CAD, edit module is the mother of all modules. It is used for creating new images or editing scanned images. The images can be loaded for editing. Edit module combines an excellent collection of painting tools and powerful retouching capabilities. The different modules of textile CAD are as follows:

- Dobby module

- Jacquard module

- Print module for printing industry

- Weave library

- Colour library

- Yarn library

5.6 WEAVE AND FABRIC PROPERTIES

Mechanical properties of fabrics are significantly influenced by the weave. Tear strength of fabrics depends on the ease of grouping of yarns during tearing. Therefore, the weaves which have long floats show higher tearing strength than the weaves which have short floats as the grouping of yarns becomes easier in the case of the former. Therefore, a five-end satin will have higher tearing strength than its equivalent 3/1 twill weave. Similarly, a 3/1 twill weave will have higher tearing strength than its equivalent 2/1 twill weave. Plain weave is expected to have minimum tearing strength.

The tensile strength of fabrics is also influenced by the weave, though the relationship is quite complex. Higher float length favours better sharing of load by the yarns due to lower obliquity effect. Plain weave has the minimum float length and the maximum crimp, which are detrimental to fabric strength. However, fabric assistance is expected to be more in the case of plain weave. The extent of crimp removal of load-bearing yarns during tensile testing has a great bearing on the tensile strength of plain woven fabrics. If the load-bearing yarns decrimp completely during tensile testing, then plain woven fabrics may show higher tensile strength than the other weaves.

Shear resistance of plain woven fabric is the maximum as it has the maximum number of interlacement. During the shear deformation, the yarns slide over each other at the crossover points. Therefore, the number of interlacements shows good association with the shear resistance of fabric.

REFERENCES

Grosicki, Z. J. 1997. *Watson's Textile Design and Colour.* Cambridge, UK: Woodhead Publishing Limited.

Robinson, A. T. C. and Marks, R. 1967. *Woven Cloth Construction.* Manchester, UK: The Textile Institute.

CHAPTER 6

Shedding

6.1 PRIMARY AND SECONDARY MOTIONS

Three primary motions are required in a loom for weaving:

1. Shedding

2. Picking

3. Beat-up

Two secondary motions are required in a loom for weaving:

1. Take-up

2. Let-off

The objective of shedding motion is to divide the warp sheet into two parts so that sufficient gap is created between them for the uninterrupted passage of the weft carrier or weft yarn from one side of the loom to the other. Picking is the operation to insert the pick (weft) from one side of the loom to the other. In shuttle looms, picking is done from both sides of the loom. However, in shuttleless looms, it is done from one side of the loom (generally from left-hand side). Several systems are available for picking. Shuttle is the most traditional system for picking and is still being used in the industry. However, productivity of shuttle loom is quite low largely due to the limitation in terms of the number of picks that can be inserted within a given time (picks per minute). In shuttleless looms, the following picking systems are used:

- Projectile

- Rapier

- Air-jet

- Water-jet

Beat-up is the operation to position the newly inserted pick up to the cloth fell, that is the boundary up to which the fabric has been woven. Take-up motion ensure the winding of fabric on cloth roller intermittently or continuously. Ensuring uniform pick spacing is also another function of take-up motion. When the fabric is wound by the take-up system, tension in the warp increases and thus it is required to release the warp from the weaver's beam, which is performed by the let-off motion.

6.2 TRANSMISSION OF MOTIONS IN SHUTTLE LOOM

The transmission of motions to some of the important loom components is shown in Figure 6.1 (Banerjee, 2015). The loom pulley (machine pulley) gets motion directly from the motor pulley. The crankshaft, which has a special design, is connected to the loom pulley. The revolution per minute (rpm) of the crankshaft is equal to the loom speed (the number of picks inserted per minute or picks/minute). Beat-up is done by the reed which is carried by the sley. Sley is a horizontal beam supported near its two ends by two upright frames called sley swords. Sley swords are mounted on a rocking shaft which positioned close to the ground level. Sley swords are connected to the crankshaft through two crank arms. One revolution of crankshaft causes one beat-up. Therefore, if 200 picks are inserted per minute, 200 beat-ups are required in 1 minute. Thus the rpm of the crankshaft has to be 200.

Crankshaft is connected to the bottom shaft through gears. As the name implies, the bottom shaft is positioned at the lower side of loom near the floor. Picking motion is originated from the bottom shaft in the case of shuttle looms. Two picking cams are mounted on the bottom shaft, one on each side. Therefore, one revolution of the bottom shaft ensures the insertion of two picks. So, for inserting 200 picks per minute, the bottom shaft should rotate at 100 rpm Therefore, rpm of the bottom shaft is always half as compared to that of the crankshaft.

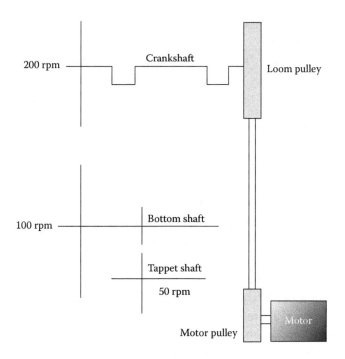

FIGURE 6.1   Transmission of motions in shuttle loom.

In the case of a plain weave, the healds return to the same position after every two picks, that is pick numbers 1, 3, 5 and so on. So the frequency of shedding operation is once per two picks. If the loom is running at 200 picks per minute, then a heald must complete 100 cycles of movement in 1 minute. Therefore, two shedding cams controlling the healds can be mounted on the bottom shaft which is rotating at 100 rpm. However, for other weaves where more than two healds are required, a cam shaft (also known as tappet shaft) is used. For example, in the case of a 3 × 1 twill weave, the shedding cycle spans over four picks. In this case, four shedding cams are mounted on cam shaft, which rotates at 50 rpm. The primary motions, their frequency, controlling loom shaft and their rpm for a plain weave and a 3 × 1 twill weave are given in Tables 6.1 and 6.2, respectively.

TABLE 6.1   Frequency of Primary Motions and Controlling Shaft for a Plain Weave

Operation	Frequency of Operation	Loom Shaft	Shaft rpm
Shedding	Once/2 pick	Bottom	100
Picking	Once/2 pick	Bottom	100
Beat-up	Once/pick	Crank	200

TABLE 6.2   Frequency of Primary Motions and Controlling Shaft for a 3 × 1 Twill Weave

Operation Cycle	Frequency of Operation	Loom Shaft	Shaft rpm
Shedding	Once/4 pick	Cam/tappet	50
Picking	Once/2 pick	Bottom	100
Beat-up	Once/pick	Crank	200

6.3 CAM SHEDDING

The cross-sectional view of a plain loom with cam shedding system is shown in Figure 6.2. Two shedding cams, mounted on the bottom shaft, are controlling two healds through treadle levers. The tips of the treadle levers are connected to the healds through ropes and links. Each treadle lever carries one treadle bowl (pulley) which actually remains in contact with the corresponding shedding cam.

For plain woven fabrics, two shedding cams are positioned at 180° phase difference. Therefore, when one cam pushes a treadle bowl in the downward direction, the other cam accommodate the upward movement of the other treadle bowl. The upward movement of the heald is activated by the roller reversing mechanism, which is positioned over the loom.

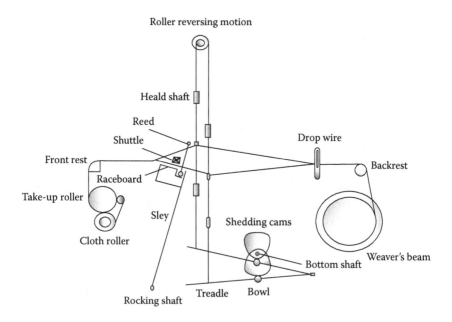

FIGURE 6.2   Cross-sectional view of loom. (From Marks, R. and Robinson, A.T., *Principles of Weaving*, The Textile Institute, Manchester, UK, 1976.)

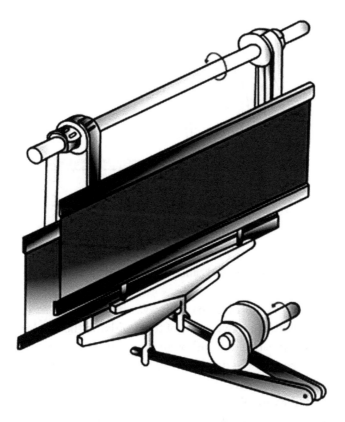

FIGURE 6.3   Cam shedding system. (Reprinted from *Weaving: Conversion of Yarn to Fabric*, 2nd edn., Lord, P.R. and Mohamed, M.H, Copyright 1982, with permission from Elsevier.)

Figure 6.3 presents another view of cam shedding system and roller reversing mechanism (Lord and Mohamed, 1982).

6.3.1 Negative and Positive Cams

The cams used for shedding can broadly be classified under two categories. In a plain loom, negative cams control only the half of the movement, that is downward movement of the healds. The upward movement is ensured by the roller or spring reversing mechanism. When back heald is pushed downward, the reversing roller rotates clockwise (Figure 6.2). This causes winding of belts, connected to the front heald, on the reversing roller having smaller diameter. Thus the front heald moves upward. Similarly, when the front heald moves downward, the reversing rollers rotate anti-clockwise. So the belt, connected to the back heald, is wound on the bigger pulley of roller reversing system. As a result the back heald moves upward.

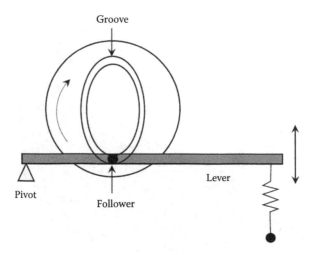

FIGURE 6.4   A model of grooved cam.

Thus, the entire system of cams and reversing mechanism controls the upward and downward movements of the healds. So the whole system can be classified as a positive one, although the cams used here are negative as they alone can control the downward movement of the healds.

Positive cams can control the upward and downward movements of the healds. Grooved cams (Figure 6.4) or matched cams are generally used as positive cams. In the case of grooved cams, the follower remains confined within a cam track. Therefore, as the cam rotates, the follower executes either vertical or lateral movement. In this case, the total movement of the follower is controlled by the cam. Therefore, no reversing mechanism is required.

6.3.2 Distinct (Clear) and Indistinct (Unclear) Shed

If the extent of vertical movement of all the healds during shedding is same, then indistinct or unclear shed is produced as depicted on the left-hand side of Figure 6.5. In the case of indistinct shed, the position of the top shed line is different for different healds. Therefore, the shuttle actually gets lower amount of space through which it has to travel. So the possibility of abrasion and collision between shuttle and shed line is higher.

In the case of distinct or clear shed, shown on the right-hand side of Figure 6.5, the position of the top shed line at the front part of the shed is same irrespective of the healds. This type of shed can be formed if the extent of vertical movement of the healds during shedding is changed based on their position. The first heald, which is nearer to the cloth fell (distance l_1), should have the minimum vertical movement and the last

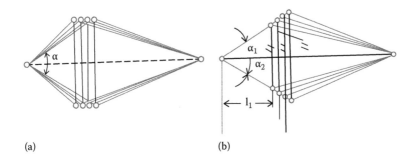

FIGURE 6.5　(a) Indistinct (unclear) and (b) distinct (clear) sheds.

heald should have the maximum vertical movement. The shuttle gets more space to travel in the case of distinct shade as $(\alpha_1 + \alpha_2) > \alpha$.

6.3.3 Lift or Throw of Cam

Lift indicates the magnitude of movement imparted by a cam to the follower. In the preceding section, we found that higher vertical movement is required for the back heald so that distinct shed is formed. However, effective length of the treadle lever is shorter for the back heald. Thus, if the lift or throw of the cams controlling the back and front heald is same, higher vertical movement will occur for the front heald. This is just opposite to the actual requirement. To overcome this problem, the cam controlling the back heald possesses a higher lift as compared to the cam controlling the front heald.

Let us consider the following parameters as shown in Figure 6.6:

x is the distance between the fulcrum point of treadle levers and centre of treadle bowl.

y is the distance between the centre of treadle bowl and tip of the treadle lever tied to the back heald.

b is the distance between the front and back healds.

a is the distance between cloth fell and front heald.

h_1 and h_2 are the lifts of the front and back healds, respectively.

L_1 and L_2 are the throws of cams controlling front and back healds, respectively.

The lift of a heald is equal to the movement of the tip of the corresponding treadle lever. Now, applying the concept of similar triangles, as shown in Figure 6.7, the following expressions can be written:

$$\frac{h_1}{L_1} = \frac{x + y + b}{x}$$

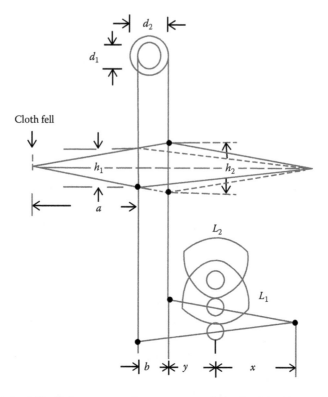

FIGURE 6.6 Lift of the cams and movement of the healds. (From Marks, R. and Robinson, A.T., *Principles of Weaving*, The Textile Institute, Manchester, UK, 1976.)

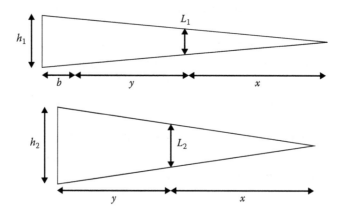

FIGURE 6.7 Schematic representation of lift of the healds and throw of cams.

so

$$L_1 = \frac{h_1 x}{x + y + b}$$

and

$$\frac{h_2}{L_2} = \frac{x + y}{x}$$

so

$$L_2 = \frac{h_2 x}{x + y}$$

Therefore,

$$\frac{L_2}{L_1} = \frac{(x + y + b)}{(x + y)} \times \frac{h_2}{h_1} \qquad (6.1)$$

At the front side of the shed,

$$\frac{h_2}{h_1} = \frac{a + b}{a} \qquad (6.2)$$

So

$$\frac{L_2}{L_1} = \left(\frac{x + y + b}{x + y} \right) \times \left(\frac{a + b}{a} \right) \qquad (6.3)$$

The right-hand side of Equation 6.3 has two components and in both the components, the numerator is greater than the denominator. Therefore, the value of L_2 is significantly greater than that of L_1. This implies that lift of the cam controlling the back heald is significantly greater than that of the cam controlling the front heald.

6.3.4 Diameter of Reversing Rollers

The shaft carrying the reversing rollers moves clockwise and anti-clockwise to control the heald movement. The angular movement of the shaft during shedding is constant. However, it has to ensure that the back heald gets higher vertical movement than the front heald so that distinct shed is produced. This is attained using two reversing rollers having different diameters. The roller with bigger diameter is connected to the back heald and vice versa. As the linear movement of a reversing roller (angular movement × radius) is equal to the vertical movement of the corresponding heald, the following expressions can be written:

$$\frac{h_2}{h_1} = \frac{d_2}{d_1} = \frac{(a+b)}{a} \tag{6.4}$$

or

$$\frac{L_2}{L_1} = \left(\frac{x+y+b}{x+y}\right) \times \left(\frac{d_2}{d_1}\right) \tag{6.5}$$

6.3.5 Geometry of Shed

A simplified geometry of shed is shown in Figure 6.8. The main shed parameters are as follows:

L_1 is length of the front shed

L_2 is length of the back shed

H is shed height

When the healds are on the warp line, that is healds are levelled, the path taken by the warp is the shortest. However, as the healds move away from the warp line, the warp takes a longer path. Thus, warp yarns are extended

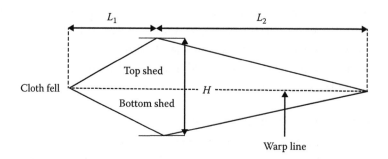

FIGURE 6.8 Geometry of shed.

which should be compensated either by the extensibility of the warp yarns or by the regulation of the let-off system. If length of the back shed is more, then yarn extension is reduced, and this is preferred for weaving delicate yarns like silk. However, shorter back shed creates clearer shed, and it is preferred for weaving coarser and hairy yarns. It is important to understand the factors which influence the degree of warp yarn extension during shed formation.

6.3.6 Calculation of Warp Strain during Shedding

A simplified mathematical model is presented here to relate the warp strain with various shed parameters. Figure 6.9 depicts a simplified view of warp strain during shedding.

Let us consider h as half of the shed height. Therefore, $H = 2h$

Elongation in the front shed = E_1.

Now,

$$E_1 = AD - AC$$

$$= \left(L_1^2 + h^2 \right)^{\frac{1}{2}} - L_1 = L_1 \left[1 + \left(\frac{h}{L_1} \right)^2 \right]^{\frac{1}{2}} - L_1$$

$$= L_1 \left[1 + \frac{1}{2} \times \left(\frac{h}{L_1} \right)^2 + \frac{\frac{1}{2}\left(\frac{1}{2} - 1\right)}{2} \left(\frac{h}{L_1} \right)^4 + \cdots \right] - L_1$$

$$= \frac{h^2}{2L_1} \tag{6.6}$$

The higher powers of h/L_1, which is <1, can be neglected.

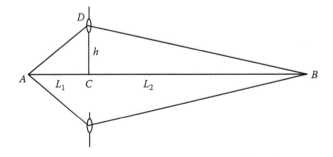

FIGURE 6.9 A simplified view of warp strain during shedding.

The ratio of lengths of front and back shed is called shed symmetry parameter (i).

Therefore,

$$\frac{L_1}{L_2} = i = \text{shed symmetry parameter}$$

Initial length of warp $= L = AB = L_1 + L_2$

$$= L_1 + \frac{L_1}{i}$$

$$= L_1\left(\frac{1+i}{i}\right) \tag{6.7}$$

Total elongation $= E = E_1 + E_2$

$$= \frac{h^2}{2L_1} + \frac{h^2}{2L_2} = \frac{h^2}{2L_1}(1+i) \tag{6.8}$$

$$\text{Strain in warp} = \frac{\text{Elongation in warp}}{\text{Initial length of warp}} = \frac{E}{L}$$

$$= \frac{1}{L} \times \frac{h^2}{2L_1}(1+i)$$

$$= \frac{h^2}{2L^2} \times \frac{(1+i)^2}{i} \tag{6.9}$$

From this equation, the following things can be inferred:

- Warp strain increases with increase in shed height (H).
- Warp strain reduces with increase in shed length (L).
- Warp strain reduces as the shed becomes symmetric (the value of i increases).

The effects of shed height and shed length on warp yarn strain are shown in Figure 6.10.

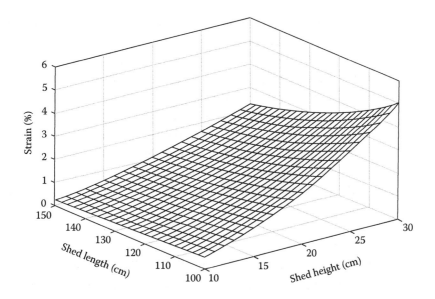

FIGURE 6.10   Warp strain during shedding.

6.3.7 Timing of Shedding

One pick cycle is equivalent to one complete rotation (360°) of the crankshaft. The timings of various loom operations are indicated corresponding to the angular position of the crankshaft which is shown in Figure 6.11.

When the crank points towards the front side of the loom (where the weaver stands), it is considered as 0° position of the crankshaft. It is also known as front centre. When the crank points downward, it is considered as 90° position of the crankshaft. It is also termed bottom centre. When the crank points towards the back side of the loom (towards the back-rest), it is considered as 180° position of the crankshaft. It is termed back centre.

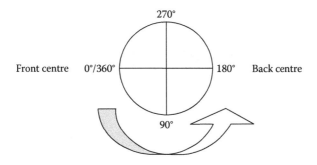

FIGURE 6.11   Different angular positions of the crankshaft.

When the crank points upward, it is considered as 270° position of the crankshaft and is termed top centre.

At 0°, the reed reaches the most forward position and performs the beat-up. On the other hand, at 180°, the reed moves to the most backward position. The sley along with the reed moves forward and backward continuously, during the entire 360° movement of the crankshaft. However, the healds do not move continuously. When the shed is completely open, the healds remain stationary for a certain time so that the shuttle can easily pass through the shed without any interference. This is called the 'dwell' period of shed. Two types of shed timing (early and late) are generally used in shuttle looms.

6.3.7.1 Early Shedding

The timing for early shedding is depicted in Figure 6.12. *E* and *L* represent the timing of shuttle entry and exit, respectively. The shuttle enters and leaves the shed at around 110° and 240°, respectively. The shed is levelled (closed) at 270°. Then it starts to open as the two healds start to move in opposite directions. The shed is fully open at 30°. Two healds are at the two extreme positions at this moment (one heald is at its top most position and another is at its bottommost position). From 30° to 150°, the healds remain stationary. Therefore, the shed is fully open and at dwell during this period. After 150°, the healds start to move in opposite

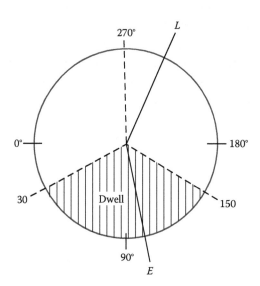

FIGURE 6.12   Timing for early shedding.

directions as compared to the movements they had between 270° and 30°. This means that the heald, which was at its topmost position, starts to descend and vice versa. The shed is again levelled at 270°.

It is understood that when the shuttle enters the shed (110°), more than half of the dwell period is over. When the shuttle leaves the shed (240°), the shed is about to close. Therefore, there is high probability that the shuttle will abrade the warp sheet, which is not desirable especially for delicate warp yarns. However, this type of timing is advantageous for weaving heavy fabric because during beat-up (0°), the shed is crossed. Therefore, the newly inserted pick will be trapped by the crossed warp yarns. As a result, the pick will not be able to move away from the cloth fell even after the reed recedes. This facilitates attaining higher pick density (picks per cm), which is required for heavy fabric.

6.3.7.2 Late Shedding

The problem of abrasion between the warp and the shuttle can be minimised by adopting late shedding, as shown in Figure 6.13. In this case, the timing of shedding is delayed in such a way that the dwell period almost coincides with the timing of shuttle flight. The shed is levelled (closed) at 0°. Then it starts to open as the two healds move in opposite directions. The shed is fully open at 120°. From 120° to 240°, the healds are stationary. Therefore, the shed is fully open and at dwell during this period. The timing of shuttle flight (110°–240°) almost coincides with the dwell period.

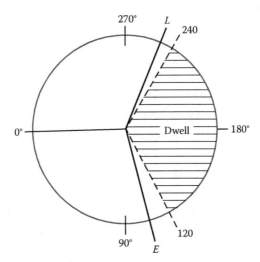

FIGURE 6.13   Timing for late shedding.

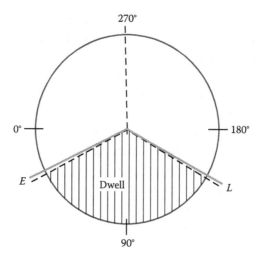

FIGURE 6.14   Infeasible adjustment of picking timing.

After 240°, the healds start to move in opposite directions and the shed is again levelled at 0°/360°.

The beat-up occurs when the shed is levelled and healds are yet to cross each other. Therefore, late shedding is not favourable for weaving heavy fabrics. However, this kind of timing is advantageous for weaving delicate warp yarns as the possibility of abrasion with the shuttle is very low (Lord and Mohamed, 1982).

One may argue that the mismatch between timings of dwell and shuttle flight, as observed in the case of early shedding, can be countered by shifting the timing of latter. So the shuttle could enter and leave the shed at around 30° and 150°, respectively, as shown in Figure 6.14. However, this change is practically not feasible as at 30° the reed will not move backward sufficiently to allow the entry of the shuttle inside the shed. It can be recalled that at 0° the reed is at the front centre and it must move backward sufficiently to create enough space for the shuttle to enter inside the shed.

6.3.7.3 Effects of Shed Timing and Backrest Position

The early shedding, coupled with raised position of backrest, results in higher pick density in woven fabric. Figure 6.15 shows the normal and raised positions of backrest. When backrest is at normal position, the top and bottom sheds are symmetrical with respect to the line CN which represents the warp line when the shed is levelled. In this case, the lengths of two shed lines – CUN and CDN – are equal, which signifies that

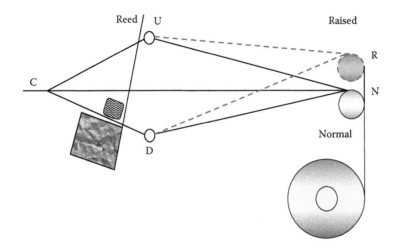

FIGURE 6.15   Warp line with normal and raised backrest positions.

the tension in warp yarns in both the sheds (top and bottom) is equal. However, when the backrest is raised from its normal position, the length of shed lines becomes unequal. This is clearly visible from the fact that the length of the top shed line CUR is smaller than that of the bottom shed line CDR. Thus the tension in the warp yarns in the top shed line will be lower than that of the bottom shed line.

In the case of early shedding, the shed is levelled at 270°. At beat-up (360°), the shed is fully crossed, that is the top shed line of the last pick has now formed a bottom shed line and vice versa. Thus the higher tension prevailing in the warp yarns of bottom shed will force the newly inserted weft (circle) in the downward direction from the fabric plane as shown at the top of Figure 6.16. This will be facilitated by the greater curvature attained by the warp yarns of top shed, which is now under low tension. The previous pick (the second circle from the right-hand side) will be forced in the upward direction with respect to the fabric plane but by a lesser magnitude. This process will repeat after the insertion of each and every pick, and as a consequence, higher pick density will result in the fabric. As the beat-up is performed at crossed shed, the newly inserted pick remains tightly meshed between the ends as the reed pushes the former towards the cloth fell against the yarn-to-yarn frictional and bending resistances. This is shown at the bottom of Figure 6.16. Once the beating is completed, the sley starts its movement towards the back centre of the loom. However, the newly inserted pick cannot spring back away from the cloth fell as it is trapped in the crossed shed.

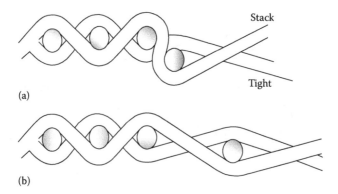

(a)

(b)

FIGURE 6.16 Vertical displacement of newly inserted weft (a) and beat-up at crossed shed (b). (From Marks, R. and Robinson, A.T., *Principles of Weaving*, The Textile Institute, Manchester, UK, 1976.)

6.3.8 Bending Factor

Bending factor is defined as the ratio of depth of shed (shed height) in front of the shuttle (s) and the actual height of the shuttle (h) as shown in Figure 6.17.

So

$$\text{Bending factor} = \frac{s}{h}$$

If it is greater than 1, then there will be no abrasion between the top warp sheet and the shuttle. Conversely, if it is much lower than 1, then severe

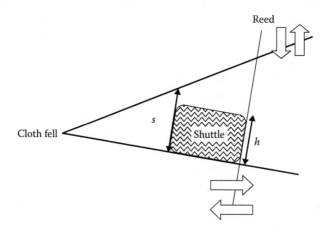

FIGURE 6.17 Bending factor.

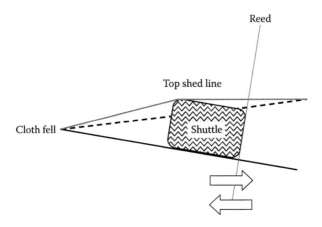

FIGURE 6.18 Situation with low bending factor.

abrasion will take place between the top warp sheet and the shuttle. The deflected top shed line is indicated by the solid line in Figure 6.18. This may lead to high warp breakage rate and even the trapping of shuttle within the shed. The bending factor changes continuously as it is influenced by the following two factors:

1. Movement of the healds

2. Movement of the sley

The bending factor will reduce as the top shed line will move in the downward direction, causing reduction in the value of s and vice versa. Besides, as the reed moves forward towards the cloth fell, the depth of shed in front of the shuttle (s) reduces. This happens because the shuttle moves over the wooden raceboard, which is carried by the sley. Thus the bending factor reduces. The reed moves towards the back of the loom between 0° and 180°. Then it moves towards the front of the loom between 180° and 360°. Thus the depth of the shed (s) changes continuously.

For late shedding, where the shed levels at 0°, dwell occurs between 120° and 240°. Therefore, during this period, the healds are stationary. So the depth of shed in front of the shuttle varies only due to the sley movement. As the sley moves to the back centre at 180°, the depth of shed becomes the maximum at this point as shown by line B in Figure 6.19. However, the reed moves forward after 180° and thus the depth of the shed reduces. After 240°, the shed starts to close and the sley is still moving forward. Both the factors synergistically reduce the depth of shed at a faster rate.

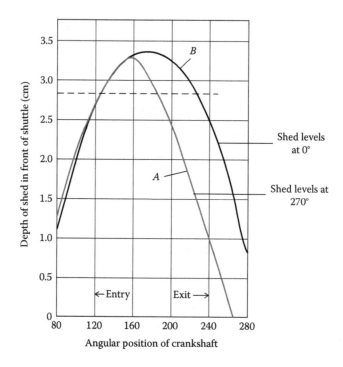

FIGURE 6.19 Bending factors for early and late shedding. (From Marks, R. and Robinson, A.T., *Principles of Weaving*, The Textile Institute, Manchester, UK, 1976.)

The actual shuttle height (2.8 cm) is indicated by the broken horizontal line. It is observed that the depth of shed is very close (slightly less) than the shuttle height when the shuttle enters and leaves the shed. So the bending factor is close to 1.

For early shedding where the shed levels at 270°, dwell occurs between 30° and 150°. After 150°, the shed starts to close. But the sley moves backward till 180°. Therefore, between 150° and 180°, two factors have a conflicting effect on the bending factor. The maximum shed height is obtained around 160° as indicated by line A in Figure 6.19. After 180°, the sley starts to move forward and the shed is still closing (till 270°). Therefore, the depth of shed reduces very fast after 180°. At the time of shuttle exit, that is 240°, the depth of shed is around 1 cm, which is just about one-third of the shuttle height. Thus, severe abrasion between the warp sheet and the shuttle is quite obvious while the latter leaves the shed. Table 6.3 shows the values of the bending factor for early and late shedding.

TABLE 6.3   Bending Factors for Early and Late Shedding

| Healds Crossing Time | Bending Factors | |
	Shuttle entering	Shuttle leaving
270°	0.87	0.34
Curve A:early shedding		
0°	0.84	0.9
Curve B: late shedding		

6.3.9 Heald Staggering

Heald staggering is done to reduce the abrasion between the warp yarns when the healds are crossing each other. When the ends per cm value in the warp sheet is very high, it is pragmatic to use four or more healds even for plain weave. If the fabric width is 200 cm and end/cm value in the fabric is 40, then total number of ends in the warp sheet is 8000. If four healds are used, then a single heald will be controlling 2000 ends. In the case of straight draft, four cams will be required to control four healds. At a particular instance, two cams will raise two healds, whereas two other cams will lower the remaining two healds. Now, at the middle of shed depth, four healds will cross each other at the same time. This will happen if four cams are paired in two groups such that there is no phase difference between the two cams of same group, whereas the phase difference between cams belonging to the two different groups is 180° on the bottom shaft. If shedding cam profile is drawn following simple harmonic motion (SHM), then velocity of the healds at the middle of the shed depth will be the maximum. Thus, when all the 8000 ends are crossing each other, enormous amount of yarn-to-yarn friction will be created, which may lead to end breakage. This can be prevented by heald staggering which ensures that all the ends are not crossing each other at the same time.

Figure 6.20 shows that the two cams of the same group (black and grey) are arranged in such a manner that there is some phase difference (say 5°–10°) between them when they are mounted on the bottom shaft. The other two cams, belonging to the other group, are at 180° phase difference with respect to the respective cams of the first group. This ensures that even when two healds are rising, they do not reach the middle of shed depth at the same time as shown in Figure 6.21. The solid and broken lines represent the movement pattern of two healds controlled by black and grey cams. Therefore, the number of ends which

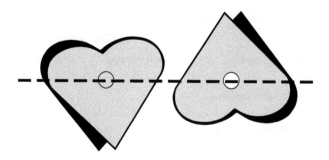

FIGURE 6.20 Schematic arrangement of shedding cams for heald staggering.

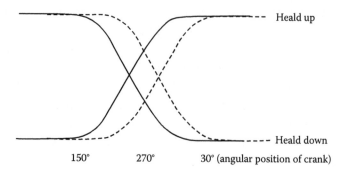

FIGURE 6.21 Crossing of healds with staggering arrangement.

cross each other at a moment is reduced, and thus the abrasion between the warp yarns is reduced considerably.

6.3.10 Heald Reversing Mechanism

When two healds are being used for weaving, a simple roller reversing mechanism can be used for raising the heald. However, a spring reversing system as shown in Figure 6.22 can also be used. In this system, a spring is extended when the healds are lowered. When the distance between the centre of bottom shaft and the periphery of shedding cam, at the point of contact with the treadle bowl, starts to decrease, action of the spring raises the healds against the gravity through connections H. This system has some drawback and can be understood from Figure 6.23.

When a heald is lowered, the warp sheet is extended. So tension (T) generates in both the limbs of warp sheet. The direction of the tension is indicated by arrowheads. The vertical components of warp tension partially compensate the weight of the heald frame (W). Therefore, the force

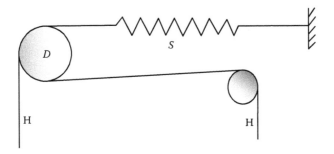

FIGURE 6.22 Spring reversing system.

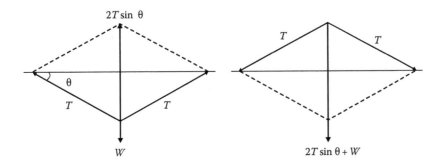

FIGURE 6.23 Forces acting on heald frames.

required to raise a heald from its bottommost position is $(W - 2T \sin \theta)$. Here θ is the angle between the bottom shed line and the horizontal plane. On the other hand, when the heald is raised, the vertical components of warp tension acts in the downward direction, which is added to the weight of the heald frame. So the force required to hold a heald at its topmost position is $(W + 2T \sin \theta)$, which is higher than the force required to raise the heald. However, the spring is fully stretched when the heald is at its bottommost position and thus maximum force is exerted on the heald frames. Conversely, the spring is least stretched when the heald is at its topmost position and minimum force is exerted on the heald frames. Thus, there is an imbalance of required force and applied force. The problem can be minimised using a specially designed pulley as shown in Figure 6.24.

As the heald is lowered, the pulley rotates anti-clockwise. Thus the effective radius of the pulley (D), on which the spring tension (S) acts tangentially, reduces. However, the radius of the pulley in the region where

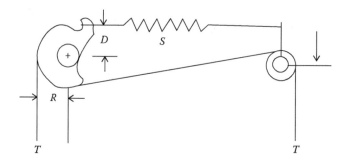

FIGURE 6.24 Modified spring reversing system. (From Marks, R. and Robinson, A.T., *Principles of Weaving*, The Textile Institute, Manchester, UK, 1976.)

the rope, connected to the heald, passes is constant (*R*). Balancing the couple, the following expression can be written:

$$2TR = S \times D$$

$$T = \frac{S \times D}{2R} \tag{6.10}$$

As the heald is lowered, *D* reduces and *S* increases. So their product remains somewhat constant. Thus, the force (*T*) acting on the heald frame connections also remains constant.

6.3.11 Positive Cam Shedding

Two types of systems are available for positive cam shedding which ensures controlled raising and lowering of healds. These are as follows:

1. Grooved cam

2. Matched cam

6.3.11.1 Grooved Positive Cam

No heald reversing motion is required in the case of positive cam shedding. A grooved cam system is shown in Figure 6.25.

The grooved cam track is formed on a circular disc. The bowl or follower is attached at one end of the quadrant (tappet lever). When the cam rotates, the bowl moves upward and downward, and this movement is translated into sidewise movement of the lower end of the tappet lever. The heald is raised and lowered using levers and link systems.

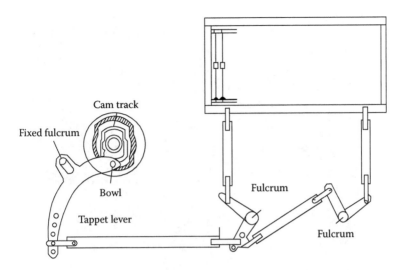

FIGURE 6.25 Grooved cam shedding system. (From Marks, R. and Robinson, A.T., *Principles of Weaving*, The Textile Institute, Manchester, UK, 1976.)

6.3.11.2 Matched Positive Cam

In matched cam systems, two cams and two followers are required for controlling a single heald. In Figure 6.26, two followers (shaded and unshaded) are attached on the two sides of the lever *L* which has fulcrum at point *F*. The shaded follower is controlled by the shaded cam and the unshaded follower is controlled by the unshaded cam. When the shaded follower is touching the minimum radius of the corresponding cam, the

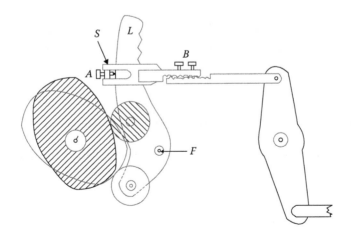

FIGURE 6.26 Matched cam shedding system. (From Marks, R. and Robinson, A.T., *Principles of Weaving*, The Textile Institute, Manchester, UK, 1976.)

FIGURE 6.27   Matched cam shedding system in a projectile loom.

unshaded follower is touching the maximum radius of the corresponding cam. The situation will be reversed when the cam shaft will rotate by 90°. This implies when the shaded follower will be pushed towards the right-hand side, the unshaded follower will be accommodated upward by the corresponding cam. Thus the lever (L) will rotate clockwise about fulcrum point F. This will cause the lifting of the heald through the connections. The matched cam system in a projectile loom is shown in Figure 6.27.

6.3.12 Design of Shedding Cams

The shedding cam has to be designed in accordance with the interlacement pattern or weave of fabric (plain, twill, satin, etc.). The design of the shedding cam influences the following:

- Dwell time of shed

- The movement pattern of the heald during raising and lowering

Shedding cams are mounted either on the bottom shaft (for plain weave and its derivatives) or on the tappet (cam) shaft. The following parameters and information are needed to design a shedding cam:

1. Weave design of the fabric

2. Minimum distance between cam and follower centres, that is nearest point of contact

3. Lift or throw of the cam (difference between maximum and minimum radius of cam)

4. Diameter of follower

5. Dwell period (duration of two dwells)

6. Duration of rise and fall of follower

7. Type of movement (linear, SHM, etc.)

It is important to remember that the number of shedding cams, with the exception of skip draft, is equal to the number of healds. The number of healds is equal to the number of ends in the repeat of weave design.

Therefore, the number of ends in the repeat of weave design determines the number of shedding cams required.

Now, one revolution of shedding cam implies n number of picks, where n is the number of picks in the repeat of weave design. Because, after n number of picks, a particular heald has to come back to the same position. Therefore, the segment of the cam available for one pick is dependent on the number of picks in the repeat of weave design.

Thus, the number of picks in the repeat of weave design determines the design of the shedding cams.

6.3.12.1 Design of Linear Cam
Example 1: Design of linear cam for plain weave
Given parameters are as follows:

- Weave: plain

- Minimum distance between cam and follower centres (d): 4 units

- Lift (l): 6 units

- Diameter of follower (f): 2 units

- Dwell period: 2/3 of a pick

- Movement pattern of follower during rise and fall: linear

Generally the duration of dwell is considered to be 1/3 of a pick. However, for weaving delicate yarns, larger dwell can be used to prevent the abrasion between the shuttle and the warp.

Rise of follower implies that the centre-to-centre distance between the cam and the follower is increasing and vice versa. The linear rise (or fall) implies that the movement of the follower in the vertical direction per unit time (per degree of cam rotation) is constant.

As it is a plain weave, 360° rotation of the bottom shaft or shedding cam correspond to two picks. Therefore, one pick is equivalent to 180° rotation of the cam. Each of the two dwells will be spanning over 2/3 × 180° = 120°. Therefore, the spans for rise and fall will be 60° each. So the movement pattern of follower can be represented as shown in Figure 6.28.

It is very impotent to note here that when the follower is rising (0°–60°), that is the distance between the cam and follower centres is increasing, the heald is actually lowering and vice versa. This happens due to the arrangement of the shedding cam and the follower in loom. When the cam radius increases, it presses the follower and the treadle lever in the downward direction. As a result the heald is lowered.

Therefore, dwell 1 and dwell 2 represent the dwell at lowered and raised positions of the shed, respectively.

Steps for drawing cam profile:

1. Draw a circle having its centre at O and radius OA of 4 units (cm or inch) as shown in Figure 6.29.

2. Add d and l (4 + 6 = 10). Then draw another concentric circle having radius of OB (10 units).

3. Divide the circle in four segments of 60°, 120°, 60° and 120°, for rise, dwell 1, fall and dwell 2, respectively. Here ∠COB = ∠DOE = 60°.

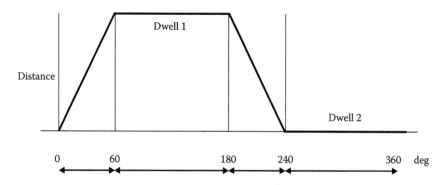

FIGURE 6.28 Linear movement pattern for a plain weave.

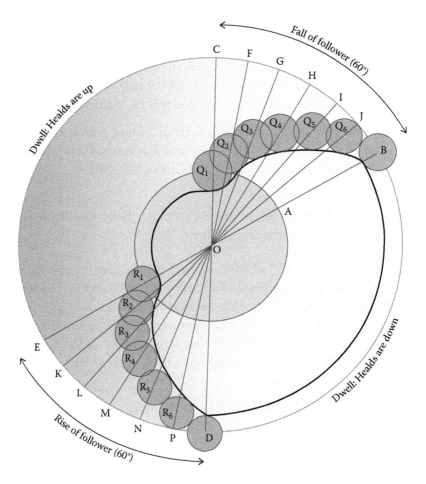

FIGURE 6.29 Linear cam for a plain weave.

4. Divide ∠COB in six equal parts by the radii OF, OG, OH, OI and OJ.
   Similarly, divide ∠DOE in to six equal parts by radii OK, OL, OM,
   ON and OP. Therefore, angles COF, FOG, GOH, HOI, IOJ and JOB
   are all equal to 10° each. Similarly, angles EOK, KOL, LOM, MON,
   NOP and POD are all equal to 10° each. Total rise of the follower is
   6 units. As the spans of rise and fall (60° for each) has been divided
   into six equal parts, that is 10° each, the distance between the centres
   of cam and follower would increase by 1/6 × 6 unit = 1 unit after each
   10° rotation of the cam during the rise of follower. During the fall it
   will be just the opposite.

5. Five arcs are drawn having centre at O and radius of 5 (4 + 1), 6 (5 + 1), 7 (6 + 1), 8 (7 + 1), 9 (8 + 1) units. The arc having radius of 5 units will cut radii OF and OK at points Q_2 and R_2, respectively. The arc having radius of 6 units will cut radii OG and OL at points Q_3 and R_3, respectively. This process will be continued.

6. Draw small circles having diameter of 2 units, representing the follower, considering Q_1, Q_2, ..., Q_6 and B as centres. Also draw small circles, having diameter of 2 units, considering R_1, R_2, ..., R_6 and D as centres.

7. Join the inner surfaces of these 14 circles with smooth curved line to get the profile of the cam (Figure 6.29).

As the cam rotation is clockwise, segment EOD causes the rise of the follower or lowering of the heald in a linear pattern. Therefore, the segment DOB (120°) causes dwell 1 of the heald at the lowered position. The segment BOC causes the fall of the follower or raising of the heald in a linear pattern. The segment COE (120°) causes the dwell 2 of the heald at the raised position. Both the dwells are equal in this case as the cam has been designed for plain weave.

Example 2: Design of linear cam for twill wave

Given parameters are as follows:

- Weave: 2/1 twill

- Minimum distance between cam and follower centres (*d*): 4 units

- Lift (*l*): 6 units

- Diameter of follower (*f*): 2 units

- Minimum dwell period: 1/3 of a pick

- Movement pattern during rise and fall of follower: linear

As it is a 2/1 twill weave, the 360° rotation of the cam shaft or cam correspond to three picks. Therefore, one pick is equivalent to 120° rotation of the cam. The two dwells, in this case, will not be equal. When the heald remains at raised position, two picks are inserted. When the heald remains at lowered position, one pick is inserted. Therefore, the dwell of heald at raised position will be longer than that of lowered position. The shorter

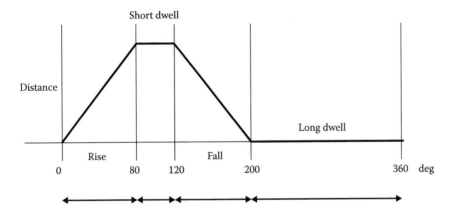

FIGURE 6.30   Linear movement pattern for a 2/1 twill weave.

dwell (at lowered position) will be spanning over $1/3 \times 120° = 40°$. Therefore, the duration of dwell in the raised position will be = duration for one pick + duration of short dwell = $120° + 40° = 160°$. Two dwells have now consumed 200° and the remaining 160° will be equally shared between rise and fall. Therefore, the spans for rise and fall will be 80° for each. So movement pattern of the follower can be represented as shown in Figure 6.30.

Steps for drawing cam profile:

1. Draw a circle having its centre at O and radius OA of 4 units (cm or inch) as shown in Figure 6.31.

2. Add d and l ($4 + 6 = 10$). Then draw another concentric circle having radius of OB (10 units).

3. Divide the circle in four segments of 80°, 40°, 80° and 160° for rise, short dwell, fall and long dwell, respectively. Here angles BOC, BOD, DOE and EOC are = 160°, 80°, 40° and 80°, respectively. Segments BOC and DOE represent the dwells of the heald at raised and lowered positions, respectively.

4. Segment COE is divided into four equal parts by the radii OK, OJ and OI. Similarly, segment BOD is divided into four equal parts by radii OF, OG and OH. Therefore, angles EOI, IOJ, JOK and COK are all equal to 20°. Similarly angles BOH, HOG, GOF and DOF are all equal to 20°. Total rise of the follower is 6 units. As the spans of rise and fall (80° each) have been divided into four equal parts, that

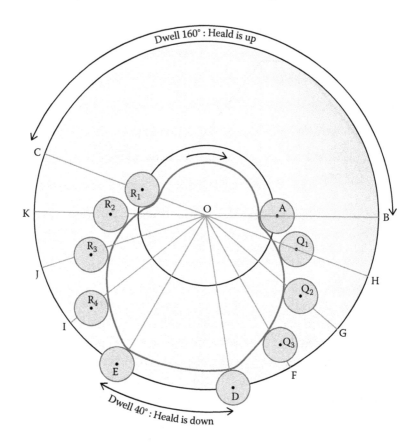

FIGURE 6.31   Linear cam for a 2/1 twill weave.

is 20° each, the distance between the centres of cam and follower would increase by ¼ × 6 unit = 1.5 unit after each 20° rotation of the cam during the rise of follower. During the fall it will be just opposite.

5. Three arcs are drawn having centre at O and radius of 5.5 (4 + 1.5), 7.0 (5.5 + 1.5) and 8.5 (7.0 + 1.5) units. The arc having radius 5.5 units will cut radii OH and OK at points Q_1 and R_2, respectively. The arc having radius 7.0 units will cut radii OG and OJ at points Q_2 and R_3, respectively. The arc having radius 8.5 units will cut radii OF and OI at points Q_3 and R_4, respectively.

6. Small circles having diameter of 2 units, representing the follower, are drawn considering R_1, R_2, R_3, R_4 and E as centres. Small circles,

having diameter of 2 units, are also drawn considering A, Q_1, Q_2, Q_3 and D as centres.

7. The inner surfaces of these 10 circles are joined with a smooth curved line to get the profile of the cam (Figure 6.31).

As the cam rotation is clockwise, segment COE causes the rise of the follower or lowering of the heald in a linear pattern. Therefore, the segment DOE (40°) causes short dwell of the heald at the lowered position. Segment BOD causes the fall of the follower or raising of the heald in a linear pattern. The segment BOC (160°) causes the long dwell of the heald at the raised position. This dwell has longer duration which is good enough for the picking system to insert two picks.

6.3.12.2 Design of Simple Harmonic Motion (SHM) Cam
Example 3: Design of SHM cam for plain weave
Given parameters are as follows:

- Weave: Plain

- Minimum distance between cam and follower centres (d): 3 units

- Lift (l): 6 units

- Diameter of follower (f): 2 units

- Dwell period: 1/3 of a pick

- Movement pattern during rise and fall of follower: simple harmonic motion (SHM)

The follower will rise and fall following SHM, which implies that movement of the follower per unit time (per degree of cam rotation) will not be constant as it happens in the case of linear movement pattern.

As it is a plain weave, the 360° rotation of the bottom shaft or shedding cam correspond to two picks. Therefore, one pick is equivalent to 180° rotation of the cam. Each of the two dwells will be spanning over $1/3 \times 180° = 60°$. Therefore, the spans for rise and fall will be 120° each. So the movement pattern of follower can be represented as shown in Figure 6.32. The rise and fall of the follower in SHM is indicated by the bold line. The broken line represents the linear movement pattern which has been given just for comparison.

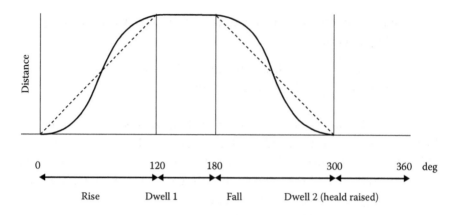

FIGURE 6.32   SHM movement pattern for a plain weave.

Steps for drawing SHM cam profile:

1. Draw a circle having its centre at O and radius (OA) of 3 units (cm or inch) as shown in Figure 6.33.

2. Add d and l (3 + 6 = 9). Then draw another concentric circle having radius (OB) of 9 units. Divide the circle into four segments of 120°, 60°, 120° and 60° for rise, dwell 1, fall and dwell 2, respectively. Here $\angle COE = \angle BOD = 120°$. The segments COE and BOD are being assigned for the fall and rise of the follower, respectively. Thus, segments BOC and DOE correspond to the dwells at lowered and raised positions of the heald, respectively.

3. Divide $\angle COE$ into six equal parts by the radii OQ_1, OR_1, OS_1, OT_1 and OU_1. Similarly, divide $\angle BOD$ into six equal parts by radii OQ_2, OR_2, OS_2, OT_2 and OU_2.

Total rise of the follower is 6 units. This is indicated by the distance FC. This implies that when the cam rotates by 120°, the follower will fall (or rise) by 6 units following SHM. In the case of SHM, a generating point has a circular motion with constant angular velocity on the circle of reference. If perpendicular lines are drawn from the generating point on the diameter of the circle of reference, then the locus of that point follows SHM.

4. To implement this, draw one semicircle considering CF as the diameter. G is the centre of the semicircle. Divide this semicircle into

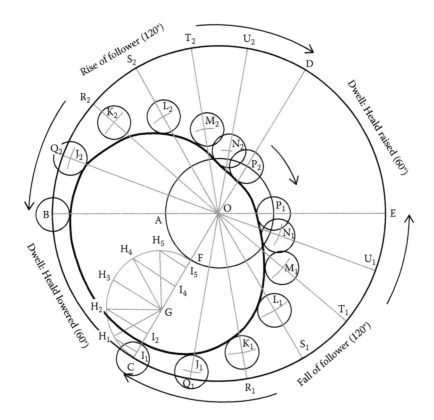

FIGURE 6.33   SHM cam for a plain weave.

six equal segments (30° each). Therefore, angles CGH_1, H_1GH_2, H_2GH_3, H_3GH_4, H_4GH_5 and H_5GF are all equal to 30° each. Draw perpendicular lines on diameter CF from points H_1, H_2, H_3, H_4 and H_5 (H_1I_1, H_2I_2, H_3G, H_4I_4 and H_5I_5, respectively).

5. Take OI_1 as radius and draw arcs which cut the radii OQ_1 and OQ_2 at J_1 and J_2, respectively. Then take OI_2 as radius and draw arcs which cut the radii OR_1 and OR_2 at K_1 and K_2, respectively. Continue these steps and in the last step take OI_5 as the radius and draw arcs which cut the radii OU_1 and OU_2 at N_1 and N_2, respectively. Now, the positions of centre of the follower have been identified during its rise and fall. These are C, J_1, K_1, L_1, M_1, N_1 and P_1, while the follower is having a fall. On the other hand, B, J_2, K_2, L_2, M_2, N_2 and P_2 are the positions of centre of the follower when it is rising.

6. Draw small circles having diameter of 2 units, representing the follower, considering the 14 centres mentioned earlier (Figure 6.33).

7. Join the inner surfaces of these 14 circles with smooth curved line to get the profile of the cam.

As the cam rotation is clockwise, segment BOD causes the rise of the follower or lowering of heald in SHM. Then, the segment BOC (60°) causes dwell 1 of the heald at lowered position. The segment COE (120°) causes the fall of the follower or raising of the heald in SHM. Segment DOE (60°) causes the dwell 2 of the heald at the raised position. Both the dwells are equal (60° or 1/3 of pick) in this case as the cam has been designed for a plain weave.

In example 3, diameter of the semicircle having centre at G is 6 units as shown in Figure 6.34. If r is the radius of the semicircle, then $2r = 6$ or $r = 3$ units.

Starting at point C (Figure 6.33) where one of the dwells is finishing, after 20° rotation of the cam (or 30° in semicircle of Figure 6.34), the displacement of the follower will be equal to the distance of CI_1. Similarly, the total displacement of the follower after 40°, 60°, 80°, 100° and 120° rotation of the cam will be given by the distances CI_2, CG, CI_4, CI_5 and CF, respectively. The expressions for these distances are shown in Figure 6.34.

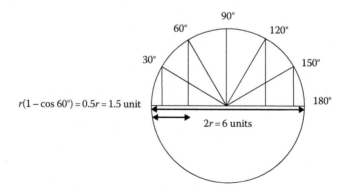

FIGURE 6.34 Displacement of the follower in SHM.

$$CI_1 = r(1 - \cos 30°) = 0.134r = 0.4 \text{ unit}$$

$$CI_2 = r(1 - \cos 60°) = 0.5r = 1.5 \text{ unit}$$

$$CG = r(1 - \cos 90°) = r = 3 \text{ unit}$$

$$CI_4 = r(1 - \cos 120°) = 1.5r = 4.5 \text{ unit}$$

$$CI_5 = r(1 - \cos 150°) = 1.866r = 5.6 \text{ unit}$$

$$CF = r(1 - \cos 180°) = 2r = 6 \text{ unit}$$

Therefore, the displacement of the follower between 30° and 60° is = $CI_2 - CI_1 = 1.5 - 0.4 = 1.1$ units. Here 30° and 60° refer to the segments of the semicircle of Figure 6.34. It should not be confused with the rotation of the cam or cam shaft. In this example, the rotation of the cam for the rise or fall of the follower is 120°. This 120° has been converted to 180° using the semicircle in Figure 6.34 so that the SHM profile can be drawn. Table 6.4 summarises the displacements according to linear and SHM movement patterns.

From Table 6.4, it is observed that the displacement made during each 30° is equal (1 unit) in the case of linear profile. In contrast, for SHM, the displacement is very small (0.4 unit) near the start and end of the motion. However, the displacement made during subsequent 30° duration increases gradually from the starting point and reaches maximum (1.5 unit) between 60° and 90°.

TABLE 6.4   Displacement during the Rise or Fall of the Follower

Span of Cam Movement (Total Span of Rise or Fall Is Equal to 180° of Semicircle)	Displacement according to Linear Profile (Unit)	Displacement according to SHM Profile (Unit)
0°–30°	1	0.4
30°–60°	1	1.1
60°–90°	1	1.5
90°–120°	1	1.5
120°–150°	1	1.1
150°–180°	1	0.4
Total	6	6

6.3.12.3 Advantages of SHM Cam over Linear Cam

If the two cams have identical design parameters except for the movement pattern (linear or SHM), then the profile of the cams will be different. If example 1 is considered then the velocity and acceleration profiles of the follower will be as shown in Figure 6.35. It is seen that between 0° and 60° and between 180° and 240°, the velocity of the follower is constant, although the direction is opposite. The constant velocity arises as the slope of the displacement profile is constant (linear line) in these two zones. Between 60° and 180° and then between 240° and 360°, the shed is at dwell and thus the velocity of the follower is zero. The follower will experience very high acceleration at 0° and 240° and very high deceleration at 60° and 180°. This can create a jerky movement as well as wear and tear of the machine components, which is highly undesirable.

Figure 6.36 depicts the velocity and acceleration profiles of the follower in the case of SHM. In this case, the change in velocity after the dwell is not sudden but gradual (as the displacement pattern is following SHM). Thus the acceleration and deceleration of the follower are much lower as compared to these of the linear movement pattern. Thus the motion of the follower is smoother and the wear and tear of the machine components is lower.

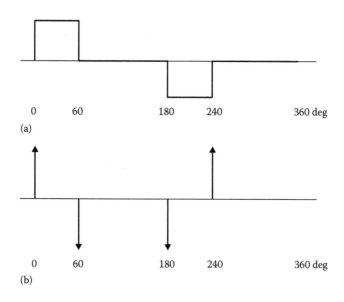

FIGURE 6.35   (a) Velocity and (b) acceleration profiles of the follower in a linear movement pattern.

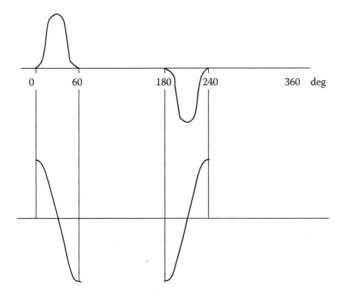

FIGURE 6.36   Velocity and acceleration profiles of the follower in SHM.

6.4 DOBBY SHEDDING

6.4.1 Limitation of Cam Shedding

The cam shedding system has limitation in terms of the number of healds that can be effectively controlled during shedding. The problem arises when the number of picks in the repeat of the weave design is very high. Let us assume that the design is repeating on 10 picks (8/2 twill). The number of healds required in this case will be 10 and, to control these healds, 10 cams will be required. These cams will be mounted on the cam (or tappet) shaft which will rotate at 1/10 rpm as compared to that of the crankshaft. Therefore, 10 picks will be inserted during one rotation of the cam shaft. As 360° rotation of the cam corresponds to 10 picks, one pick becomes equivalent to 36°. If the dwell is one-third of a pick, then the short dwell period (when heald is down) will be = 1/3 × 36° + 36° = 48°. The duration of movement of heald (when the radius of the cam changes) will be 2/3 × 36° = 24° each for upward and downward direction. This will create the long dwell (when the heald is up) of 12° + 7 × 36° = 264°.

Now, the aforementioned calculation reveals that the follower has to move from the lowest position to the highest position within 24° span. The follower must follow the contour of the cam profile which becomes steep when the span available for follower movement is low. Besides, the force

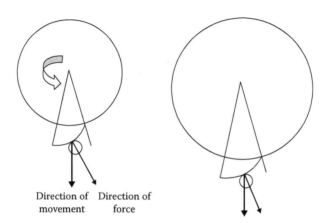

Direction of   Direction of
movement   force

FIGURE 6.37   Influence of cam dimension on steepness of cam contour.

acts on the follower in a direction which is perpendicular to the tangent drawn on the cam contour. However, the follower has to move vertically up or down. Thus, an angle is created between the direction of applied force and the direction of movement of the follower. This angle becomes higher when the cam contour is steeper, that is the span available for movement is low, which has resulted from a higher number of picks in the design repeat. This leads to the fact that only one component of the force applied by the cam becomes effective in creating the movement of follower. Thus, a very high force is actually required to create the desired movement of the follower which may lead to wear and tear as well as vibration in the system.

One plausible solution to the aforesaid problem could be to increase the dimension of the cams (Figure 6.37). It can be observed that the steepness of the cam contour, for a given span of follower movement, decreases when the diameter of the cam increases. However, this may create another problem in terms of power consumption and space availability in the system. Therefore, a dobby shedding system is preferred when a very large number of healds are to be controlled by the shedding mechanism.

6.4.2 Keighley Dobby

Keighley dobby is known to be a double acting dobby as most of the operations are done at half speed as compared to the loom speed (picks per minute). The basic components of Keighley dobby are as follows:

- Stop bars
- Baulk

- Hooks (two per heald)

- Knives (two for the entire dobby)

- Pegs on pattern chain

Figure 6.38 depicts the simplified view of Keighley dobby.

The motion of the reciprocating knives (K_1 and K_2) originates from the bottom shaft of loom. As one revolution of the bottom shaft ensures two picks, each of the two knives completes the cycle of inward (towards the left-hand side) and outward (towards the right-hand side) movements during this period. The two reciprocating knives are in complete phase difference. When one knife is moving inward, the other knife is moving outward. In Figure 6.38, knife 2 (K_2) has pulled hook 2 (H_2) towards the right-hand side. This has happened as there is a peg in the lag corresponding to the position of feeler 2 (F_2). The peg has pushed the right end of feeler 2 in the upward direction. Thus the left end of feeler 2 has

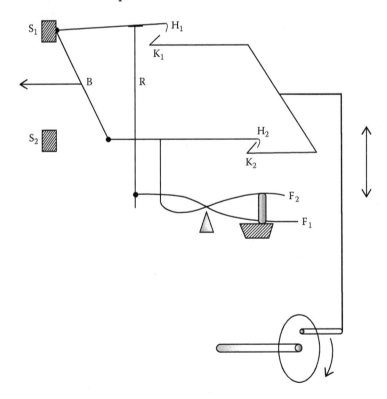

FIGURE 6.38 Keighley dobby. (From Marks, R. and Robinson, A.T., *Principles of Weaving*, The Textile Institute, Manchester, UK, 1976.)

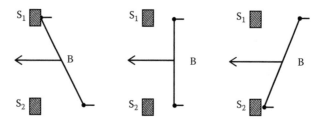

FIGURE 6.39 Open shed formation in Keighley dobby.

been lowered. So hook 2 is also lowered on knife 2 when the latter moves inward. So the lower end of baulk (B) moves away from stop bar 2 (S_2) with hook 2. Thus the heald is raised as it is connected to the midpoint of the baulk.

In the next part of the cycle, knife 2 will move inward and knife 1 (K_1) will move outward. Now, there is no peg corresponding to the position of feeler 1 (F_1). So the right end of feeler 1 is lowered due to its own weight and the left end of it is raised. As a result, connecting rod (R) has pushed hook 1 in the upward direction. So when knife 1 will perform its outward movement, it will not be able to catch hook 1. The top part of baulk will be resting on stop bar 1 and thus the heald will not be raised in the next pick.

It is important to note here that when the heald has to be in lowered position for two consecutive picks, the top as well as the bottom end of the baulk will be resting on the respective stop bars, that is S_1 and S_2. So the midpoint of the baulk will not have any significant movement. On the other hand, if the heald has to be in raised position for two consecutive picks, then one end of the baulk will move away from the stop bar and another end of the baulk will move towards the stop bar. Thus the midpoint of the baulk will not experience any significant movement as schematically shown in Figure 6.39. So the heald will remain in a raised position between the two picks. Therefore, the system will produce an open shed. Thus the amount of wasted movement is very nominal.

6.4.2.1 System of Pegging

A three up three down one up one down twill weave (3/3/1/1), which repeats on eight ends and eight picks (Figure 6.40), has been considered here for demonstrating the pegging plan. The system of pegging is depicted in Figure 6.41. This design can be produced using eight healds and straight draft. The selection for heald movement is controlled by wooden pegs which are inserted within the circular holes made on the wooden lags.

8		×				×	×	×
7	×				×	×	×	
6				×	×	×		×
5			×	×	×		×	
4		×	×	×		×		
3	×	×	×		×			
2	×	×		×				×
1	×		×				×	×
	1	2	3	4	5	6	7	8

Picks (vertical axis) — Ends (horizontal axis)

FIGURE 6.40   Point paper representation of a 3/3/1/1 twill weave.

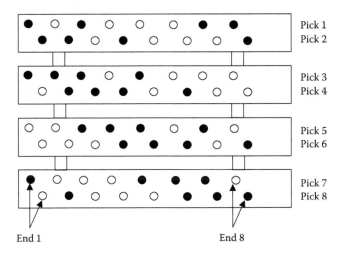

FIGURE 6.41   Peg plan for a 3/3/1/1 twill weave.

A black circle in Figure 6.41 indicates the presence of a peg. So, for the first pick, the healds 1, 3, 7 and 8 should be raised. This has been ensured by placing wooden pegs in 1st, 3rd, 7th and 8th positions corresponding to pick 1 in lag 1. The wooden lags are inked together into a lattice which is mounted on the pattern wheel (or barrel). The pattern barrel is rotated by a certain degree once in two peaks. For example, if the barrel is hexagonal then it must rotate by 60° after every two picks. The presence of a peg within a hole results in a raised position of the heald and vice versa. For the same heald, the positions of two holes for the two consecutive picks are not on the same line. The lateral shifting of holes is required so that two adjacent feelers can be accommodated.

6.4.3 Cam Dobby

In the case of cam dobby, two negative cams are used to impart motion to the knives. The cam shaft carries two cams which are positioned diametrically opposite. The cam shaft rotates at half of the loom speed. The knives are being carried by levers which are pivoted as shown in Figure 6.42. As the cam shaft rotates, one of the cam pushes its follower towards the left-hand side, causing the corresponding lever to rotate around its pivot point. At the same time, the other cam accommodates the movement of its follower towards the right-hand side which is caused by the contraction of its return spring. For example, Figure 6.42 shows that the unshaded cam has pushed its follower towards the left-hand side, causing anti-clockwise movement of the unshaded lever around its pivot point, which extends the lower return spring. At the same time, the shaded cam has accommodated the movement of the shaded follower towards the right-hand side as the centre-to-centre distance between them is the lowest at this instance. Therefore, the shaded lever also rotates anti-clockwise as the top return spring exerts pressure at the fastening point of the lever.

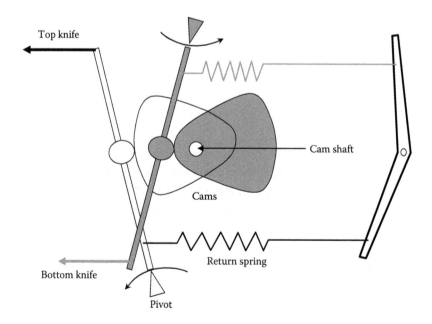

FIGURE 6.42 Cam dobby.

6.4.4 Positive Dobby

In Keighley dobby, the knives and hooks cause movement to the baulk and as a result, the heald is raised. The lowering of heald is done by the reversing system. So Keighley dobby is classified as a negative dobby. However, the upward and downward movements of the healds are completely controlled by the positive dobby. This can be achieved by the rotary dobby as shown in Figure 6.43.

Figure 6.43 depicts the simplified view of a rotary-type positive dobby. The system utilises specially designed toothed gears for causing engagement or disengagement of gears and transmission of motions. Each and every heald is controlled by a specially designed toothed gear B, which can be rotated either by the gear A or by the gear C which have teeth only over the half of their periphery. However, the gear B can mesh only with one of the driver gears (A or C) at a time. The gears A and C complete one revolution in every pick. As they rotate in opposite directions, they can rotate the gear B in anti-clockwise or in clockwise directions which is required to lower or raise the heald shaft, respectively, through the links.

The selection mechanism presents cylinders of different diameters for different picks. If the diameter of the cylinder is low, then the lever carrying the gear B is lowered on the gear C as shown in Figure 6.43. A missing tooth on the gear B facilitates the meshing between the two gears. The gear C now rotates the gear B in the anti-clockwise direction by half revolution, causing the heald to be lowered through links. The heald will

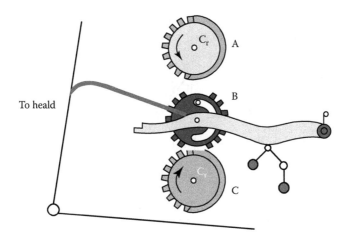

FIGURE 6.43 Positive dobby. (From Marks, R. and Robinson, A.T., *Principles of Weaving*, The Textile Institute, Manchester, UK, 1976.)

retain its lowered position as long as cylinders with lower diameter will be presented by the selection mechanism to the lever. This happens because the gear B has a portion where three teeth are missing and it is now at the zone of contact between gears B and C. Therefore, meshing is not possible and no further rotation is caused to the gear B, though the gear C is continuously rotating in every pick. If a cylinder with higher diameter is fed by the selection mechanism, then the lever will be raised and thus the gear B will move in the upward direction to mesh with gear A. The single missing tooth of the gear B, which is now at the zone of contact between gears A and B, will again facilitate the meshing gears. The gear B will now rotate in clockwise direction, causing the heald to be raised.

6.4.5 Modern Rotary Dobby

Rotary dobby converts the rotational movement into linear movement, which is required for lifting and lowering of healds. Rotary dobby can operate at a very high speed, up to 1500 rpm. The operating principle of Fimtextile RD 3000 rotary dobby is shown in Figure 6.44. The cam shaft rotates by 180° and then stops momentarily, and thus the motion is called irregular rotary motion. The cam unit is mounted on the cam shaft but

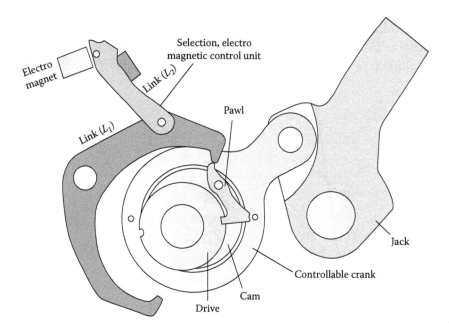

FIGURE 6.44 Fimtextile RD 3000 rotary dobby. (Courtesy of Fimtextile Rotary Dobby RD3000, Fimtextile S.p.A, Italy.)

not fixed on it. The pawl, which is placed on the outside of the cam, can connect it with the driver and then the cam rotates by 180°, causing the movement to the heald. The crank unit encloses the cam with ball bearings. Link (L_1) can rotate around its pivot by the action of the electromagnet through link L_2. If link L_1 rotates in an anti-clockwise direction, then the pawl rotates in clockwise direction and its bottom tip engages with the groove on drive. If link L_1 rotates in a clockwise direction, it presses the upper tip of the pawl and thus disengages it from drive. When the engagement happens, the jack rotates in an anti-clockwise direction during the 180° rotation of the dobby shaft. The jack dwells at its foremost position during the 180° rotation of the dobby shaft, if the engagement does not occur. The dobby shaft stops after every 180° degree rotation and the pattern selection mechanism engages or disengages the pawl with the drive.

6.5 JACQUARD SHEDDING

Jacquard shedding system was developed by Joseph Marie Jacquard (1752–1834), who was a French weaver and merchant. In the case of cam and dobby shedding systems, a large number of warp yarns passing through a heald are controlled in a group. Thus, it precludes the possibility of controlling individual ends independently. Therefore, complicated woven designs cannot be made using cam or dobby shedding systems. With Jacquard shedding systems, individual ends can be controlled independently and thus large woven figures can be produced on the fabrics.

Mechanical Jacquard systems can be classified under three categories:

1. Single lift and single cylinder (SLSC)

2. Double lift and single cylinder (DLSC)

3. Double lift and double cylinder (DLDC)

6.5.1 Single-Lift Single-Cylinder (SLSC) Jacquard

Figure 6.45 shows the simplified side view of SLSC Jacquard. Single-lift implies that one end is controlled by one hook which is the lifting element. If the Jacquard has a capacity to handle 300 ends independently, then it requires 300 hooks (one per end) which are arranged vertically and 300 needles (one per hook) which are arranged horizontally. For example, the needles can be arranged in six rows and each row will have 50 needles. In the side view, only six needles (one per horizontal row) are visible. Hooks, which are connected to individual ends through nylon

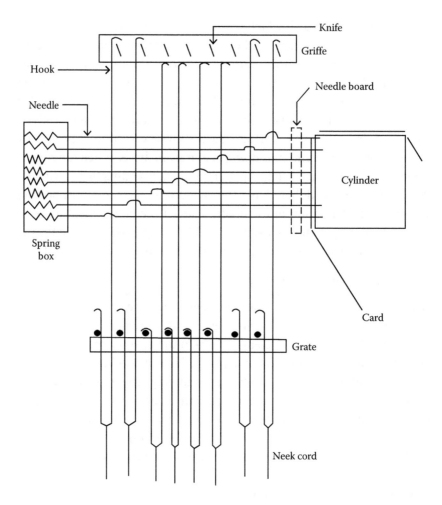

FIGURE 6.45 Side view of single-lift single-cylinder Jacquard. (From Marks, R. and Robinson, A.T., *Principles of Weaving*, The Textile Institute, Manchester, UK, 1976.)

cord (harness), are also arranged in six rows each having 50 hooks. One knife is responsible for controlling the movement (lifting and lowering) of one row of hooks. However, whether a hook will be lifted or not will be ascertained by the selection mechanism which is basically a punched card system mounted on a revolving cylinder having square or hexagonal cross-section. The needles are connected to springs at the opposite side of cylinder. Therefore, the needles always exert some pressure on the right-hand side (Figure 6.45). So, if there is a hole in the punch card corresponding to the position of a needle, then the needle will be able to pass through the hole and thus the corresponding hook will remain in upright position, making it accessible to the knife when the latter has started its upward movement after descending to the lowest height. On the other hand, if there is no hole, then the needle will be pressed towards the left-hand side against the spring pressure. Thus the kink, which partially circumscribes the stem of a hook, present in the needle presses the hook towards the left-hand side making the latter tilted enough from the vertical plane so that the knife misses it while moving upward. Therefore, presence of a hole implies selection (ends up) and vice versa. A hole in this case is tantamount to a peg used on the lag of dobby shedding system.

In the case of SLSC Jacquard, if the loom speed is 300 picks per minute, the cylinder will turn 300 times per minute (5 times per second) and the knives should also reciprocate (up and down) 300 times per minute. Thus, it hinders the high loom speed. When a particular hook and the corresponding end have to be in a raised position in two consecutive picks, the former descends to its lowest possible height (determined by the grate) in between the two peaks and then moves up again. Thus, it produces a bottom closed shed. This happens as one end is controlled by a single hook.

Features of SLSC Jacquard are as follows:

- 500 end machine will have 500 needles and 500 hooks.

- Cylinder should turn in every pick.

- Knives must complete the cycle of rise and fall in every pick.

- Bottom closed shed is produced.

6.5.2 Double-Lift Single-Cylinder (DLSC) Jacquard

Double-lift implies that one end is controlled by two hooks. Double-lift single-cylinder (DLSC) Jacquard is shown in Figure 6.46. In this case, two hooks which control a single end are again controlled by a single needle. For example, hooks 1 and 2 control end 1 and hooks 3 and 4 control end 2. Two sets of knives are used in DLSC Jacquard and they move up and down (rise and fall) in complete phase difference, that is when one set of knives (K_1 and K_3) attain the highest position, the other set of knives (K_2 and K_4) attain the lowest position and vice versa. At the given position, end 1 has been raised as hook 1 has been lifted by the corresponding knife K_1. However, end 2 has not been raised as hook 3 was not caught by knife K_3. In the next pick, end 1 will be lowered as the

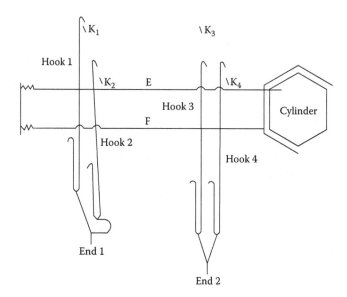

FIGURE 6.46 Side view of double-lift single-cylinder Jacquard. (From Marks, R. and Robinson, A.T., *Principles of Weaving*, The Textile Institute, Manchester, UK, 1976.)

needle F has been pressed towards the left-hand side due to the absence of a hole in the punch card. So hook 2 has become tilted and will not be lifted by knife K_2 when the latter will rise. Hook 1 will also descend with knife K_1. Thus, end 1 will be lowered in the next pick as both the hooks (1 and 2) will be in lowered position. On the other hand, end 2 will be raised in the next pick as there is a hole in the punch card corresponding to the position of the needle E. So hook 4 is upright and will be caught by knife K_4 when the latter will move upward.

In the case of DLSC Jacquard, if the loom speed is 300 picks per minute then the cylinder will turn 300 times per minute but the knives will reciprocate (rise and fall) 150 times per minute. This is the advantage of DLSC Jacquard over SLSC Jacquard.

DLDC Jacquard produces semi-open shed because if a particular end has to be in a raised position for two consecutive picks, it will descend up to the middle point of its vertical path and then move up. This will happen because one of the hooks will descend and the other hook will move up with their respective knives and they will cross at the middle of their vertical path. If the end has to remain in the lowered position for two consecutive picks, it will remain so without any intermediate movement.

The features of DLSC Jacquard are as follows:

- 500 end machine will have 500 needles and 1000 hooks.

- Two sets of knives rise and fall in opposite phase.

- Cycles of movement (rise and fall) of each set of knives spans over two picks.

- Cylinder should turn in every pick.

- Semi-open shed is produced.

6.5.3 Double-Lift Double-Cylinder (DLDC) Jacquard

Figure 6.47 depicts the double-lift double-cylinder (DLDC) Jacquard. In the case of (DLDC) Jacquard, the number of cylinder rotation or turn and the number of reciprocation cycle of knives per minute are half as compared to those of SLSC. In this case, one end is controlled by two hooks as it was in the case of DLSC. However, each of the hooks is controlled by separate needles. Hooks 1 and 2 control end 1, and hooks 3 and 4 control end 2. Needles 1, 2, 3, and 4 control hooks 1, 2, 3 and 4, respectively.

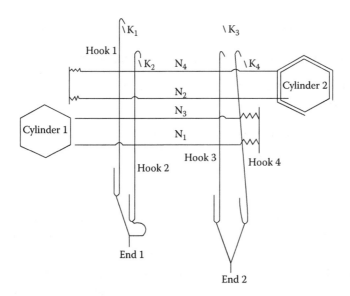

FIGURE 6.47 Side view of double-lift double-cylinder Jacquard. (From Marks, R. and Robinson, A.T., *Principles of Weaving*, The Textile Institute, Manchester, UK, 1976.)

The two needles (say N_1 and N_2) corresponding to a particular end (say end 1) are controlled by two cylinders in two picks. One of the needles (N_2) is controlled by the right cylinder (cylinder 2) and the other needle (N_1) is controlled by the left-hand side cylinder (cylinder 1). One cylinder carries the punch cards for even-numbered picks N, N + 2, N + 4, N + 6 and so on. Here N is an even number. The other cylinder carries the punch cards for odd-numbered picks N + 1, N + 3, N + 5 and so on. In one pick, either of the two cylinders performs the selection operation. DLDC Jacquard is capable of handling the maximum loom speed (picks per minute) among the three types of Jacquard.

Figure 6.47 shows that end 1 is in a raised position and end 2 is in the lowered position in this current pick. End 1 will continue to be in a raised position in the next pick as there is a hole in punch card on cylinder 2 corresponding to the position of needle 2 (N_2). So hook 2 will remain in an upright position and thus it will be raised by knife 2 (K_2). On the other hand, end 2 will continue to be in the lowered position as it is being tilted by needle 4 (N_4) as there is no hole on cylinder 2 corresponding to the position of N_4. So knife 4 (K_4) will miss the N_4 when the former will rise in the next pick.

Features of DLDC Jacquard are as follows:

- 500 end machine will have 1000 needles and 1000 hooks.

- Two sets of knives rise and fall in opposite phase.

- Cycles of movement (rise and fall) of each set of knives spans over two picks.

- Cylinder should turn in alternate pick.

- Semi open shed is produced.

6.5.4 Jacquard Harness

Jacquard harness is the system by which the ends are controlled during Jacquard shedding with the help of nylon cords, heddles (elements like heald eyes) and dead weights (lingoes). In the preceding part of discussion, we considered the capacity of Jacquard to be 300 hooks. Now, if the fabric has 3000 ends then 10 repeats of the design can be produced on the fabric. For example, if a floral pattern is woven on the fabric and the pattern requires 300 ends, then 10 such floral patterns can be produced on the entire width of the fabric. It is assumed that the hooks of the Jacquard are arranged in six rows and each row has 50 hooks. Then each hook will effectively control 10 (3000/300) ends. The interlacement pattern of ends 1, 301, 601, 901, 1201, 1501, …, 2701 will be identical, and thus, they can be controlled by hook 1 through 10 nylon cords (harness cords). Similarly, hook 300 will control 10 ends, namely ends 300, 600, 900, 1200, 1500, 1800, …, 3000. This is depicted in Figure 6.48. The individual harness cords pass through the perforations of a wooden or polymer board named comber board. The dead weights or lingoes pull the end downward when it is not lifted. Harnessing is a very time-consuming activity and it is seldom changed.

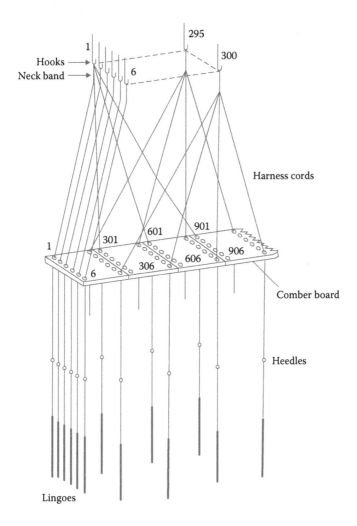

FIGURE 6.48 Jacquard harness. (From Marks, R. and Robinson, A.T., *Principles of Weaving*, The Textile Institute, Manchester, UK, 1976.)

6.5.5 Problems in Jacquard Harness in Wide Looms

When the width of loom and comber board is high, problem may arise in terms of variation in the lift received by various heddles and consequently the ends they control. This will be clear from Figure 6.49. Here the loom width is 200 cm and the perpendicular distance between the neck band suspended from a hook and the comber board is 150 cm when the end is down. The lift of the hook is 10 cm.

Therefore, when the hook is raised by 10 cm, then the heddle, which is exactly at the bottom of the hook, will receive a lift of 10 cm. However, the heddle which is at the extreme left or right of the comber board will receive a lift equal to the following:

$$
\begin{aligned}
P'R - PR &= \sqrt{(L+\Delta L)^2 + \left(\frac{W}{2}\right)^2} - \sqrt{(L)^2 + \left(\frac{W}{2}\right)^2} \\
&= \sqrt{(150+10)^2 + \left(\frac{200}{2}\right)^2} - \sqrt{(150)^2 + \left(\frac{200}{2}\right)^2} \\
&= \sqrt{(160)^2 + (100)^2} - \sqrt{(150)^2 + (100)^2} \\
&= 188.7 - 180.3 = 8.4 \text{ cm}
\end{aligned}
\tag{6.11}
$$

where
  L is the perpendicular distance between neck band and comber board
  ΔL is the lift of the hook
  W is the width of the comber board

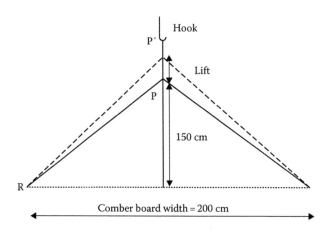

FIGURE 6.49 Variation in Jacquard lift.

Thus the loss of lift is around 16% with respect to the end positioned vertically below the hook.

Now, if the height of the Jacquard is increased by 100 cm, then the perpendicular distance between the neck band suspended from a hook and the comber board would become 150 cm + 100 cm = 250 cm. In this case, the heddle, which is at the extreme left or right of the comber board, will receive a lift as shown as follows:

$$= \sqrt{(250+10)^2 + \left(\frac{200}{2}\right)^2} - \sqrt{(250)^2 + \left(\frac{200}{2}\right)^2}$$

$$= \sqrt{(260)^2 + (100)^2} - \sqrt{(250)^2 + (100)^2}$$

$$= 278.57 - 269.26 = 9.31 \text{ cm}$$

Thus the loss of lift reduces from 16% to 6.9%. This is reason for mounting the Jacquard system at a sufficient height over the loom.

6.5.6 Pattern of Harness Tying

There are different ways in which harness can be tied to control the ends. Figure 6.48 shows a straight tie. This is also shown schematically in the upper part of Figure 6.50. If the Jacquard capacity is 300 hooks and the total number of ends to be controlled is 1200, then four repeats can be produced across the width of the fabric. In this case, the maximum number of ends in the repeat is equal with the Jacquard capacity. For designs which are symmetrical across the width, pointed tie can be used as shown in the

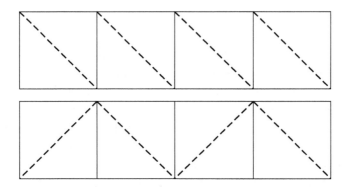

FIGURE 6.50 Straight and pointed harness tie.

FIGURE 6.51   Symmetrical design.

lower part of Figure 6.50. In this case, the maximum number of ends in the repeat becomes double the Jacquard capacity.

Figure 6.51 depicts a symmetrical heart to be woven on a fabric using a Jacquard loom. Let us assume that the design is repeating on 600 ends and the Jacquard capacity is only 300 hooks. This can be achieved using a pointed tie where ends 1 and 600 are controlled by, say, hook 1, and ends 2 and 599 are controlled by hook 2 and so on.

6.5.7 Electronic Jacquard

In recent times, electronic Jacquards have become very popular. They control the ends by synchronised operations of following machine components:

- Electromagnet

- Retaining hook or ratchet

- Hooks

- Knives

- Double roller

The knives (K_1 and K_2) are used to lift or lower the hooks (H_1 and H_2). If the electromagnet is activated by a signal, then it can briefly retain the upper end of the retaining hooks (R_1 and R_2) once the latter is pressed on the electromagnet due to the upward movement of the hook. If this happens then the hook is not retained by the retaining hook when the former starts to descend with the knife. On the other hand, if the electromagnet is not activated, then the hook is retained or caught by the retaining hook. Figure 6.52 depicts the operation of the electronic Jacquard system.

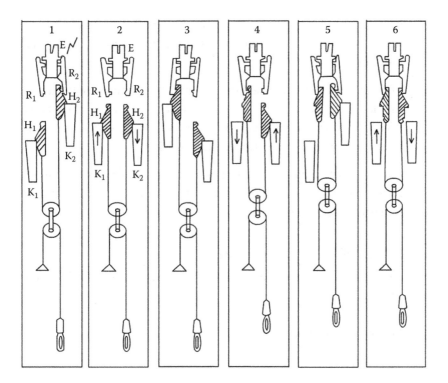

FIGURE 6.52   Principles of operation of electronic Jacquard.

Position 1: Hook 2 (H$_2$) has been lifted to the highest position by knife 2 (K$_2$). However, the electromagnet (E) has been activated and it holds the top end of the retaining hook 2 (R$_2$) momentarily and thus ensures that the retaining hook 2 does not catch hook 2 when the latter descends. At this instance the shed is at a lower position.

Position 2: Knife 2 and hook 2 are descending whereas knife 1 (K$_1$) and hook 1 (H$_1$) are moving up. So there is no effective movement of the double pulley assembly or the heddle. The shed is still at a lower position.

Position 3: Hook 1 has been raised to the highest position by knife 1 and thus hook 1 has pressed the top end of retaining hook 1 (R$_1$) against the electromagnet. At this moment the electromagnet has not been activated, which ensures that hook 1 is caught by the retaining hook 1.

Position 4: Knife 1 has started to descend but hook 1 cannot descend as it is caught by the retaining hook 1. Knife 2 has again started its upward movement with hook 2. The shed has now started to change its position (moving upward). This is because hook 1 is already in a raised position as it is caught by retaining hook 1 and hook 2 is also moving up.

Position 5: Hook 2 has pressed the top end of retaining hook 2 against the electromagnet which is not activated at this instance. Thus, hook 2 is caught by the retaining hook 2. As both the hooks are now in a raised position, it creates upper shed position.

Position 6: Knife 2 has started to descend. However, hook 2 retains its raised position as it is held by the retaining hook 2. Thus the shed remains in upper position.

6.6 TYPES OF HEALD MOVEMENT

Depending on the types of heald movement, sheds can be classified under four categories:

1. Bottom closed shed

2. Semi-open shed

3. Centre closed shed

4. Open shed

The movement pattern of healds for different sheds is shown in Figure 6.53 considering the weave to be a 3/2 twill.

6.6.1 Bottom Closed Shed

In the case of bottom closed shed, all the ends come to their lowest position after every pick to close the shed. If an end is supposed to be in raised position in two consecutive picks, then it will come to the lowered or bottom position between two picks. This movement is unnecessary and thus there is lot of wasted movement.

Example: Single lift Jacquard produces a bottom closed shed.

6.6.2 Semi-Open Shed

In the case of semi-open shed, if an end has to be in raised position in two consecutive picks, then it comes down up to the middle of shed depth (up to warp line) and then again goes up. However, if one end has to be in

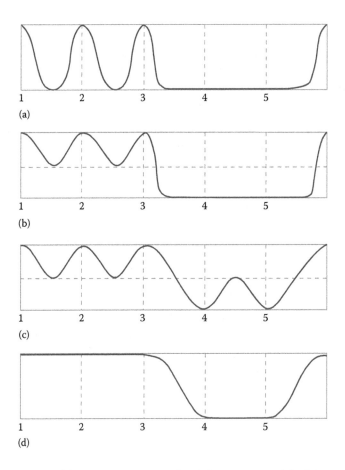

FIGURE 6.53 Heald movement pattern for different types of shed (3 up 2 down twill). (a) Bottom closed shed, (b) semi-open shed, (c) centre closed shed and (d) open shed.

lowered position in two consecutive picks, it does not move at all between picks. Overall, the wasted movement is lower for semi-open shed as compared to that of the bottom closed shed.

Example: Double lift Jacquard produces semi open shed.

6.6.3 Centre Closed Shed

In this case, the shed closes at the middle of shed depth (at the warp line) after every pick. Therefore, if an end has to be in raised (or lowered) position in two consecutive picks, it will come to the middle of shed depth between the two picks. The amount of wasted movement is also very high in this case.

6.6.4 Open Shed

This is the ideal kind of shed and it minimises the wasted movements of the ends or healds. If an end has to be in raised position in two consecutive picks, then it remains stationary in the raised position between two picks. Similarly, if the end has to be in lowered position in two consecutive picks, then it remains stationary in the lowered position between two picks.

Example: Keighley dobby and cam produce open shed.

6.7 DUAL-DIRECTIONAL SHEDDING

Fabrics used for apparel applications primarily have length and width dimensions. The thickness is negligible in most of the cases. However, in many technical applications such as composites and aerospace, a thicker and integrated fabric structure is required. 3-D weaving is one of the ways to produce sufficiently thick fabrics. True 3-D weaving can be achieved only by having a shedding system that can enable interlacing of a grid-like multiple-layer warp yarns with vertical and horizontal sets of weft to produce a fully interlaced 3-D fabric. This kind of shedding system is known as dual-directional shedding. In this case, the multi-layer warp yarns move in two directions, that is along the fabric thickness and width. This enables the formation of sheds in columns and rows as shown schematically in Figure 6.54.

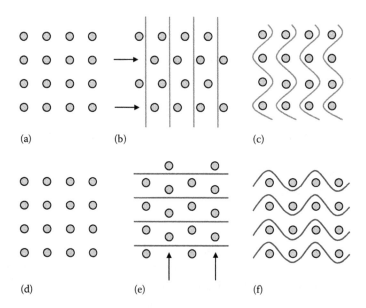

FIGURE 6.54   Principle of dual-directional shedding.

The 2nd and 4th rows of ends have moved towards the right-hand side, creating sheds in vertical direction (Figure 6.54b). Then the vertical picks are inserted. The shed closes as the 2nd and 4th rows of ends come back to the starting position, causing interlacement between warp and vertical picks. Figure 6.54e shows the situation when the 2nd and 4th columns of ends have moved upward, creating horizontal sheds. Then the horizontal picks are inserted. When the shed closes again, interlacement takes place between warp and horizontal picks. The final 3-D woven fabric is shown in Figure 6.55.

Sometimes non-interlacing-type 3-D fabrics are produced using multiple layers of warp in a 2-D loom. This technique cannot be considered as true weaving as no real shed is formed in this case. Figure 6.56 presents a method to produce non-interlacing-type 3-D fabrics in a 2-D loom. The three layers of warp yarns are being delivered from weaver's beams shown with black circles. These warp yarns, positioned along X axis, are also called stuffer yarns. The two sets of binder yarns, indicated by broken

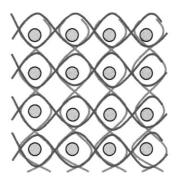

FIGURE 6.55 Interlacement in 3-D woven fabric.

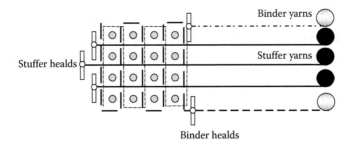

FIGURE 6.56 Non-interlacing 3-D fabric weaving.

lines, are being controlled by two separate healds and they are being delivered from two weaver's beams shown with grey circles. Binder yarns are positioned along Z axis and they run through the thickness of the fabric. The picks are indicated by smaller grey circles and are positioned along the Y axis. After the insertion of one set of picks (four in this case), the two healds controlling the binder yarns interchange their positions and thereby bind the stuffer and weft yarns. Therefore, the stuffer warp, binder warp and picks are mutually perpendicular to each other. However, there is no true interlacement in the fabric.

NUMERICAL PROBLEMS

6.1 Calculate the throw (lift) of the cam controlling the back heald from the following particulars:

Throw (lift) of the cam controlling the front heald = 8 cm

The distance between the front and back healds = 4 cm

The distance between the fulcrum and centre of bowl on the treadle = 20 cm

The distance between the centre of bowl and the fastening point of the back heald = 20 cm

Diameter of small reversing roller = 5 cm

Diameter of large reversing roller = 6 cm

Solution:

The expression to be used for calculating the throw or lift of the cam is as follows:

$$\frac{L_2}{L_1} = \left(\frac{x+y+b}{x+y} \right) \times \left(\frac{a+b}{a} \right)$$

Here

L_2 is the throw (lift) of the cam controlling the back heald

L_1 is the throw (lift) of the cam controlling the front heald = 8 cm

x is the distance between the fulcrum and centre of bowl on the treadle = 20 cm

y is the distance between the centre of bowl and the fastening point of the back heald = 20 cm

b is the distance between the front and back healds = 4 cm

d_2 is the diameter of large reversing roller = 6 cm

d_1 is the diameter of small reversing roller = 5 cm
a is the distance between cloth fell and front heald (not given)
Now,

$$\left(\frac{a+b}{a}\right) = \frac{d_2}{d_1}$$

So

$$\frac{L_2}{L_1} = \frac{x+y+b}{x+y} \times \frac{d_2}{d_1}$$

$$L_2 = L_1 \times \frac{x+y+b}{x+y} \times \frac{d_2}{d_1}$$

$$= 8 \times \frac{20+20+4}{20+20} \times \frac{6}{5} = 10.56$$

So the lift of the cam controlling the back heald is 10.56 cm.

6.2 Determine the ratio of strains created in warp yarns during shedding by the front and back healds if the total shed length (distance between the cloth fell and backrest) is 120 cm, front shed length for the front heald is 20 cm, distance between the front and back heald is 4 cm and diameters of reversing rollers are 5 and 6 cm.

Solution:
The strain in warp yarns can be expressed by the following equation:

$$\text{Strain} = \frac{h^2}{2L^2} \times \frac{(1+i)^2}{i}$$

For the front heald:
Front shed length (L_1) = 20 cm
Therefore,
The back shed length = L_2 = (120 – 20) = 100 cm
So

$$\text{Shed symmetry parameter} = i_1 = \frac{L_1}{L_2} = \frac{20}{100} = 0.2$$

For the back heald:

$$\text{Front shed length} \left(L_1'\right) = \left(20+4\right)\text{cm} = 24\text{ cm}$$

Therefore,

$$\text{The back shed length} = L_2' = \left(120-24\right) = 96\text{ cm}$$

So

$$\text{Shed symmetry parameter} = i_2 = \frac{L_1'}{L_2'} = \frac{24}{96} = 0.25$$

Now,

$$\frac{h_2}{h_1} = \frac{d_2}{d_1} = \frac{6}{5} \quad \text{and the total shed length } L$$

$$\text{is same for both the healds.}$$

So

The ratio of strains in warp yarns for front and back healds

$$= \frac{h_1^2}{h_2^2} \times \frac{\dfrac{\left(1+i_1\right)^2}{i_1}}{\dfrac{\left(1+i_2\right)^2}{i_2}} = \frac{h_1^2}{h_2^2} \times \frac{\left(1+i_1\right)^2}{\left(1+i_2\right)^2} \times \frac{i_2}{i_1}$$

$$= \frac{5^2}{6^2} \times \frac{\left(1.2\right)^2}{\left(1.25\right)^2} \times \frac{0.25}{0.2} = 0.8$$

So the ratio of strains is 0.8:1, which implies that the back heald is creating more strain in the warp yarns.

REFERENCES

Banerjee, P. K. 2015. *Principles of Fabric Formation*. Boca Raton, FL: CRC Press.
Lord, P. R. and Mohamed, M. H. 1982. *Weaving: Conversion of Yarn to Fabric*, 2nd edn. Merrow, UK: Merrow Technical Library.
Marks, R. and Robinson, A. T. C. 1976. *Principles of Weaving*. Manchester, UK: The Textile Institute.

CHAPTER 7

Picking in Shuttle Loom

7.1 OBJECTIVES

The objective of picking is to propel the weft yarn or the weft carrying element (shuttle, projectile or rapier) along the correct trajectory, maintaining requisite velocity through the shed to provide lateral sets of yarns in the fabric. In this chapter, picking will be discussed with reference to shuttle loom only.

7.2 LOOM TIMING

Loom timing is defined as relative chronological sequences of various primary and secondary motions expressed in terms of angular position of the crankshaft. The loom timing, for early shedding, is shown in Figure 7.1. The sley moves continuously either forward (180°–360°) or backward (0°–180°) as indicated by the innermost circle. However, the heald movement is not continuous due to the shed dwell as indicated by the middle circle. The outer circle shows the timing for shuttle entry and exit.

The timing of various events can be summarised as follows:

Shedding (for early shedding)

30°: Shed is fully open.

30°–150°: Shed dwell and shed remains fully open.

150°–270°: Shed is closing.

270°: Shed is closed or shed levelled.

270°–30°: Shed is opening.

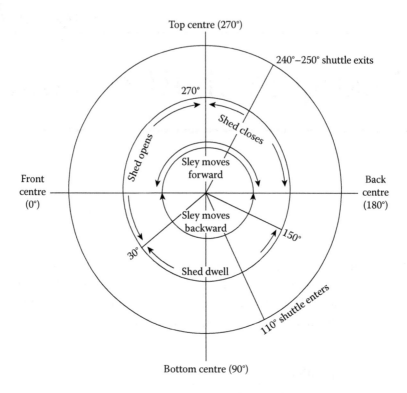

FIGURE 7.1 Timing diagram of shuttle loom for early shedding.

Shedding (for late shedding)

120°: Shed is fully open.

120°–240°: Shed dwell and shed remains fully open.

240°–360°: Shed is closing.

360°: Shed is closed or shed levelled.

360°–120°: Shed is opening.

It can be seen that all the operations have been delayed by 90° in the case of late shedding.

Picking and checking

80°–110°: Picking mechanism operates.

105°–110°: Shuttle enters the shed.

240°–250°: Shuttle leaves the shed.

270°: Shuttle strikes the swell in the shuttle box.

300°: Shuttle comes to rest.

Beat-up

0°: Beat-up takes place and sley occupies its forward most position.

0°–180°: Sley moves backward.

180°: Sley occupies its backward most position.

180°–360°: Sley moves forward.

Take-up

0°–10°: Take-up (intermittent type).

7.3 CLASSIFICATION OF SHUTTLE PICKING MECHANISM

Shuttle picking mechanisms are broadly classified as cone over-pick and cone under-pick. The distinction arises from the position of movement of the picking stick which occurs over the sley in the case of over-pick and below the sley in the case of under-pick. Modifications of cone under-pick mechanism are available as parallel pick and link pick.

7.3.1 Cone Over-Pick Mechanism

The cone-over pick mechanism is shown in Figure 7.2. A picking cam attached to the bottom shaft displaces the picking cone which is attached to the upright picking shaft. This causes rotation of the picking shaft. As a result, the picking stick, which is attached to the uppermost end of picking shaft, swings in a horizontal plane over the loom and transmits the motion to shuttle through the picking strap and the picker guided by a spindle. Picking strap is a leather or polymeric belt which is flexible. Here the picker is constrained by the spindle to move in a straight line which otherwise would have followed a path of arc. Moreover, pair of picking cams and followers (picking cones) installed at either side of the loom have seldom ensured picking of equal strength (force). The cams responsible for impulsive rotation of the picking stick receive motion through the bottom shaft. However, the allied system of picking, that is

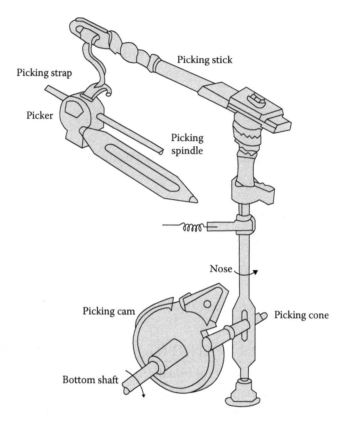

FIGURE 7.2   Cone over-pick mechanism.

picking stick, picking strap and so on, has varying elastic behaviour. All these necessitate frequent adjustment of picking-strap or picking cam and nose settings. A cone over-pick mechanism on a loom is shown in Figure 7.3.

7.3.1.1 Adjustments for Strength and Timing of Over-Pick Mechanism

- Shortening picking-strap increases the shuttle speed, but the timing of picking advances.

- The picking cam can be turned over the bottom shaft for advancing or delaying the timing of picking. If the picking cam is turned anti-clockwise (Figure 7.2), then the timing of picking will be delayed and vice versa.

FIGURE 7.3   Cone over-pick mechanism on a loom.

- Lowering the picking cone in its slot increases the shuttle speed but the timing of picking is delayed. Figure 7.4 shows the picking cone in raised and lowered positions. When picking cone is lowered on its slot, the distance between the axis of bottom shaft and the axis of picking cone reduces. So the latter receives the full throw of the cam, causing more rotation of the picking shaft and thus the shuttle speed increases. At lowered position, the picking cone comes into contact with the cam after some delay, causing the delay in the timing of picking.

- Angular adjustment between the picking shaft and the picking stick also changes the shuttle speed and timing (unpredictable).

- Large change in shuttle speed for wider loom can be achieved by changing either the nose bit or the entire picking cam.

With the advent of automatic loom which comes with magazine for holding the full pirns, replacement of over-pick system with cone under-pick

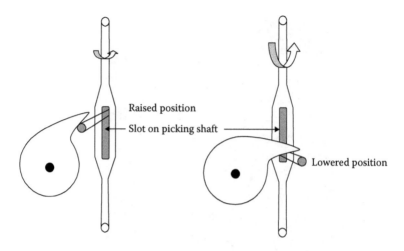

FIGURE 7.4   Change in the position of a picking cone.

has become inevitable. The cone under-pick mechanism provides space over and at one end of the loom for pirn changing mechanism.

7.3.2 Cone Under-Pick Mechanism

Cone under-pick mechanism is depicted in Figure 7.5. A picking cam attached to the bottom shaft displace the cone, turning the horizontally

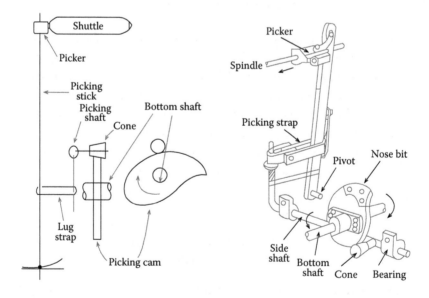

FIGURE 7.5   Cone under-pick mechanism. (From Marks, R. and Robinson, A.T., *Principles of Weaving*, The Textile Institute, Manchester, UK, 1976.)

Picker

Spindle

Lug or picking strap

Picking stick

FIGURE 7.6 Cone under-pick mechanism on a loom.

located picking shaft (side shaft). The other end of the picking shaft is connected to the upright picking stick through the picking strap or lug strap. This causes the picking stick to move in a vertical plane and transmits the motion to shuttle by the picker attached at its upper end. In this system, the picking stick and other elements are located below the shuttle trajectory. The system is naturally suitable for automatic looms. Here the picker slides over the metallic spindle and the picking timing is regulated by cam adjustment like in over-pick motion. An almost inextensible lug strap allows shuttle speed adjustment by either raising or lowering it around picking stick. Absence of stretchable parts in under-pick system ensures the retention of correct setting over long period of loom operation. A cone under-pick mechanism on a loom is shown in Figure 7.6.

7.3.2.1 Adjustments for Strength and Timing of Under-Pick Mechanism

- Timing of picking is adjusted by turning the cam on the bottom shaft.

- Raising and lowering of the lug strap (picking strap) reduces and increases shuttle velocity, respectively. This happens because the stroke or displacement of the picker increases with the lowering of lug strap.

Two independent adjustments for velocity of shuttle and timing of picking make the under-pick system less complicated.

7.3.2.2 Parallel Pick

Parallel pick is a modified under-pick mechanism in which picker guiding spindle has been eliminated. The picker itself is kept attached to the stick as shown in Figure 7.7. The lower end of the stick is made to oscillate over a curvilinear shoe which rests upon a plate. The shoe and picker trajectory form the circumference of a semicircle as shown in Figure 7.8. This ensures picker movement in a perfect straight line and hence that of the shuttle which is a prime requirement for high-speed looms. The picking stick in its extension passes through a slot of a plate and connected to loom frame through a spring for proper return (Marks and Robinson, 1976).

7.3.2.3 Link Pick

Attempt to increase loom speed further creates one undesirable situation where contact between the shoe and the plate is often lost, resulting in unwanted picker movement which deflects shuttle from its right path or undue wear of it and even fly-out. Modification through link pick is done to overcome this limitation as shown in Figure 7.9. Here a metal piece (M) is attached to the lower end of the picking stick. M connects itself to a bracket (B) fastened to the sley sword through two arms (A). Such four

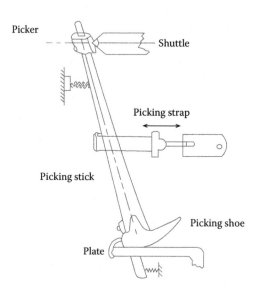

FIGURE 7.7 Parallel pick mechanism.

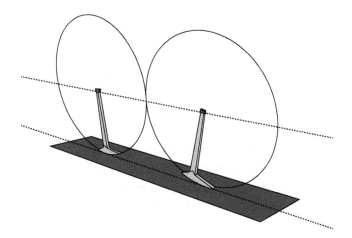

FIGURE 7.8   Principle of parallel pick mechanism.

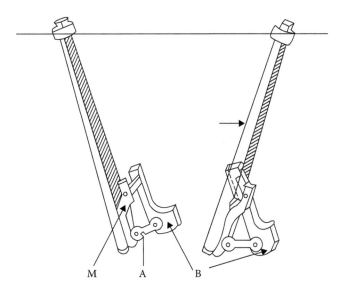

FIGURE 7.9   Link-pick mechanism. From Marks, R. and Robinson, A.T., *Principles of Weaving*, The Textile Institute, Manchester, UK, 1976.

bar linkages form an irregular quadrilateral whose shape and sizes in relation to the length and angular movement of the picking stick, if properly designed, can deliver a very good result in terms of accuracy over a distance of 16–20 cm along the shuttle path (Marks and Robinson, 1976). This system truly exercises positive control in the system even at very high speed.

7.3.2.4 Side Lever Under-Pick Mechanism

Side lever under-pick mechanism is a very simple system. Figures 7.10 and 7.11 depict the front and the side views of this mechanism, respectively. Two wooden picking levers (side levers), which are placed at the two sides of loom and run perpendicular to the bottom shaft, are used in this mechanism. The bottom shaft carries two picking discs, one at each side, which have slots for holding the picking bowls. Each of the picking discs carries one picking bowl. However, the two picking bowls are positioned at 180° phase difference so that one revolution of the bottom shaft can cause the insertion of two picks one from each side of the loom. Each side lever carry a picking shoe which has a profiled shape to ensure the desired

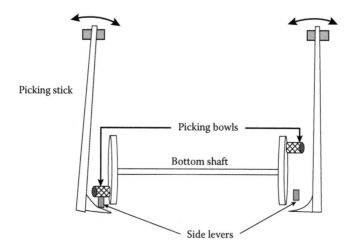

FIGURE 7.10   Front view of side lever picking mechanism.

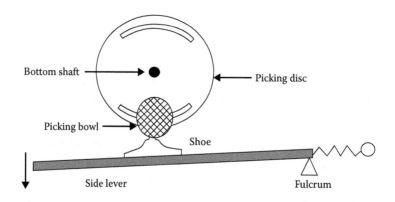

FIGURE 7.11   Side view of side lever picking mechanism.

contact with the picking bowl. When a picking bowl comes into contact with the curvature of the shoe, the side lever is depressed as depicted in Figure 7.11. Thus the side lever presses the shoe attached to the lower part of the picking stick. So the picking stick moves inwards and propels the shuttle. After the picking, the side lever and the picking stick return to their original positions due to the action of springs. The timing of the pick can be adjusted by changing the position of the picking bowl on the slot of picking disc (Banerjee, 2015).

7.4 CATAPULT EFFECT

Generally, shuttle picking mechanism operates from 80° to 110° of angular position of crankshaft. Basically, the picking duration comprises two halves. In the first half (from 80° to 95°) the strain energy is built up and in the latter half (from 95° to 110°) the strain energy is released. The shuttle picking mechanism has a close analogy with the action of a catapult. In this context, the shuttle represents the missile, the picker represents the leather part and the picking stick and the picking strap represent the elastic band. In the case of shuttle picking, the strain energy is stored in the first half by bending of picking stick, twisting of picking shaft and stretching of picking strap. In the latter half, the strain energy is released, which accelerates the shuttle. Therefore, in the light of such strong analogy the shuttle may be said to be catapulted by picking mechanism.

7.5 SHUTTLE VELOCITY, LOOM SPEED AND PICKING POWER

7.5.1 Relation between Shuttle Velocity and Loom Speed

Let us assume the following notations:

P = Loom speed (picks/min) or number of revolutions of crankshaft/min

R = Width of the reed (m)

v = Average shuttle velocity (m/s)

L = Effective length of shuttle (m)

θ = Degree of crankshaft rotation available for the travel of the shuttle through the shed

t = Time required for the travel of the shuttle through the shed (s)

Now the distance covered by the shuttle in t s $= (R + L)$ m
So

$$\text{Average shuttle velocity } v = \frac{R+L}{t} \text{ m/s}$$

or

$$t = \frac{R+L}{v} \text{ s} \tag{7.1}$$

P revolution of the crankshaft takes 60 s.
So 1 revolution of the crankshaft takes 60/P s.
Therefore, 360° rotation of the crankshaft takes 60/P s
or θ° crankshaft rotation takes place in $\theta/6P$ s.
So

$$t = \frac{\theta}{6P} \tag{7.2}$$

From Equations 7.1 and 7.2, we get

$$\frac{R+L}{v} = \frac{\theta}{6P}$$

or

$$v = \frac{6P(R+L)}{\theta} \tag{7.3}$$

or

$$P = \frac{v\theta}{6(R+L)} \tag{7.4}$$

Equation 7.4 can be interpreted as follows: If the distance travelled by the shuttle, that is $(R + L)$ is 1.5 m, and the average shuttle velocity is 15 m/s, then the shuttle takes 0.1 s to travel through the shed. If the degree of crankshaft rotation (θ) available for the passage of the shuttle through the shed is 120°, then the loom speed becomes 200 picks per minute. Because 120° rotation of the crankshaft becomes equivalent to 0.1 s, 360° rotation of the crank-shaft becomes equivalent to 0.3 s. Therefore, loom speed = 60/0.3 = 200 picks per minute. Now, if the average shuttle velocity is increased to 20 m/s, then the shuttle will take 0.075 s to travel through the shed. If θ remains unchanged, 120° rotation of the crankshaft becomes equivalent to 0.075 s. Thus the loom speed increases to 267 picks per minute. On the other hand, if the average shuttle velocity remains 15 m/s but θ increases from 120° to 150°, then 150° rotation of the crankshaft becomes equivalent to 0.1 s and so the loom speed increases to 250 picks per minute.

It can be inferred from Equation 7.4 that for a given loom width and shuttle length, to increase the loom speed, either v or θ or both should be increased. However, an increase in θ can be achieved by increasing the sley eccentricity (e) as the sley remains towards the back centre of the loom for a longer duration if sley eccentricity is more. Thus the shuttle can avail greater duration for its flight through the shed. A high value of sley eccentricity should be avoided to reduce wear and tear of the loom. In addition, an increase in θ may cause problem for fast reed warp protector motion. Because in order to prevent the shuttle trapping inside the shed, the shuttle should strike the swell in proper time. Therefore, the increase of θ is not an easy option. On the other hand, an increase in v requires more kinetic energy to be dissipated during shuttle checking, which increases wear of the loom components. From this discussion, it is clear that the loom speed is limited in the shuttle loom.

The product of P and (R + L) is known as the weft insertion rate (WIR), which is a measure of loom productivity. Unit of WIR is m/min.

From Equation 7.4, for a given v and θ, the following expression can be written:

$$P \propto \frac{1}{(R+L)} \tag{7.5}$$

Equation 7.5 indicates that loom speed is inversely proportional to (R + L). However, this theoretical relationship largely deviates from the actual one. A good approximation of the relationship between the loom speed and the loom width can be expressed by using the inverse-square-root rule as expressed by Equation 7.6.

$$P \propto \frac{1}{\sqrt{(R+L)}} \tag{7.6}$$

7.5.2 Power Required for Picking

The energy used to accelerate the shuttle is equal to its kinetic energy when it leaves the picker.

So

$$\text{Energy/pick} = \frac{1}{2}mv^2 \text{ J}$$

where

   m is the mass of the shuttle (kg)
   v is the shuttle velocity (m/s)

Now

   Power required for picking = Work done/s for picking

   = Energy used to accelerate the shuttle/s.

If the loom speed is P picks/min, then the number of picks inserted/s = $P/60$.

Therefore

   Power required for picking = $1/2\ mv^2 \times P/60 \times 1/1000$ kW.

Now from Equation 7.3

$$v = 6P(R + L)/\theta$$

By substituting the expression of v, the following relationship is obtained.

$$\text{Power required for picking} = \frac{3mP^3(R+L)^2}{\theta^2} \times 10^{-4} \text{ kW} \qquad (7.7)$$

So

$$\text{Energy/pick} = \frac{1}{2}mv^2 = \frac{1}{2}m\left(\frac{6P(R+L)}{\theta}\right)^2 = \frac{18mP^2(R+L)^2}{\theta^2} \text{ J} \qquad (7.8)$$

Equation 7.7 suggests that for a given width of loom, the power required for picking is proportional to the cube of loom speed (picks/min). Therefore, if the loom speed is increased by 20%, the power required for picking will increase by 73% (1.2^3 times = 1.728 times). However, the productivity of fabric (m/min) will also increase by 20% due to increased loom speed. So the power cost per unit length (m) of fabric will increase by 44% (1.728/1.2 times).

Power of picking is also influenced as follows:

- Increases linearly with the mass of the shuttle

- Increases proportionately with the square of loom width

- Decreases proportionately with the square of degree of crankshaft rotation available for the travel of the shuttle through the shed

7.6 NOMINAL AND ACTUAL DISPLACEMENT OF SHUTTLE

The plot of shuttle displacement versus angular position of the crankshaft is shown in Figure 7.12. If the loom is operated slowly by hand, the shuttle displacement is termed nominal displacement, which is indicated by a solid line. The shuttle displacement during picking under actual running condition of loom is known as actual displacement which is indicated by a broken line. It can be seen that these two displacements differ significantly due to the inertia effect of the shuttle. The actual displacement lags behind the nominal displacement, and the lag increases with the increase in crankshaft rotation. At a point near the halfway of picking (95°), the lag

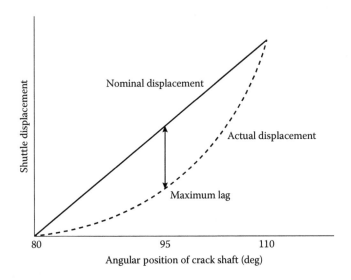

FIGURE 7.12 Nominal and actual displacement of the shuttle.

becomes maximum. The strain energy stored in the picking system is maximum at this point. Therefore maximum picking force acts on the shuttle at this point. After this, the lag between nominal and actual displacements reduces and thus the force acting on the shuttle also diminishes. The shuttle loses contact with the picker at the point where the actual and nominal displacement curves cross each other (110°). At this instance, the actual and nominal displacements are equal and therefore no further picking force acts on the shuttle. The shuttle has gained its maximum velocity and leaves the picker. This has close analogy with the catapult effect during shuttle picking as mentioned earlier.

If it is assumed that the picking stick is elastic, the force generated by the stick is directly proportional to the distance through which the stick is bending, that is $(s - x)$, where s and x are nominal and actual displacement of shuttle, respectively.

The force exerted on the picking stick = $\eta(s - x)$, where η is equivalent rigidity (force required per unit deformation) of the picking system.

Again

$$\text{The force acting on the shuttle} = m\frac{d^2x}{dt^2}$$

where

m is the shuttle mass

$\dfrac{d^2x}{dt^2}$ is the actual acceleration of shuttle

So

$$m\frac{d^2x}{dt^2} = \eta(s - x)$$

or

$$\frac{d^2x}{dt^2} = \frac{\eta}{m}(s - x)$$

or

$$\frac{d^2x}{dt^2} = n^2(s - x) \tag{7.9}$$

where n is a constant which is termed the alacrity of the picking system and its expression is given by $\sqrt{\eta/m}$.

Equation 7.9 is identical to the equation of SHM, which implies that the picking system is a vibrating one having its own particular natural frequency. The alacrity of the picking system is directly proportional to the natural frequency.

7.6.1 Nominal Movement in Straight Line

For straight-line nominal movement of shuttle displacement

$$s = k\theta$$

where
  k is the proportionality constant
  θ is the angular movement of the crankshaft

Now

$$\theta = \omega t$$

where
  ω is the angular velocity of the crankshaft
  t is the time at which the crankshaft is turned by an angle θ

So

$$s = k\omega t$$

From Equation 7.9,

$$\frac{d^2x}{dt^2} + n^2x = n^2 k\omega t$$

The solution of this differential equation is

$$x = k\omega \left(t - \frac{\sin nt}{n} \right) \qquad (7.10)$$

Equation 7.10 is the expression of actual displacement of shuttle.

Thus the actual shuttle velocity $\dfrac{dx}{dt} = k\omega(1 - \cos nt)$ and the actual acceleration of shuttle $\dfrac{d^2x}{dt} = k\omega n \sin nt$.

Now, the maximum actual velocity of shuttle is achieved when $\cos nt = -1 = \cos\pi$.

So

$$\left(\frac{dx}{dt}\right)_{max} = 2k\omega \quad \text{when } t = \frac{\pi}{n} \tag{7.11}$$

The maximum actual acceleration of shuttle is achieved when $\sin nt = 1 = \sin(\pi/2)$

So

$$\left(\frac{d^2x}{dt^2}\right)_{max} = k\omega n \quad \text{when } t = \frac{\pi}{2n} \tag{7.12}$$

Equations 7.11 and 7.12 imply that the maximum acceleration of shuttle is dependent on the loom speed and the alacrity of the picking system, whereas the maximum shuttle velocity is dependent on the loom speed only. Besides, the maximum acceleration of shuttle takes half time ($\pi/2n$) as compared to the time required for the maximum actual velocity of shuttle (π/n).

Now, at $t = \pi/n$, the maximum actual velocity of shuttle is achieved. So at $t = \pi/n$, $\left(\frac{dx}{dt}\right) \rightarrow \left(\frac{dx}{dt}\right)_{max}$.

From Equation 7.10,

$$L = \frac{\pi k\omega}{n} \tag{7.13}$$

where L is the effective stroke of the picker which is the distance moved by the shuttle at its maximum actual velocity.

Again, from Equation 7.12, we get

$$n = \frac{\left(\dfrac{d^2x}{dt^2}\right)_{max}}{k\omega} \tag{7.14}$$

By substituting the value of n in Equation 7.13, we get

$$L = \frac{\pi(k\omega)^2}{\left(\dfrac{d^2x}{dt^2}\right)_{max}} \tag{7.15}$$

or

$$\left(\frac{d^2 x}{dt^2}\right)_{max} = \frac{\pi (k\omega)^2}{L}$$

By substituting the expression of $k\omega$ from Equation 7.11, we get

$$\left(\frac{d^2 x}{dt^2}\right)_{max} = \frac{\pi}{4L}\left(\frac{dx}{dt}\right)^2_{max} \qquad (7.16)$$

Therefore, for a given maximum actual velocity, the maximum actual acceleration is inversely proportional to the effective stroke of the picker.

Now, nominal shuttle displacement is given by

$$s = k\omega t$$

So, nominal velocity of shuttle $\dfrac{ds}{dt} = k\omega$ = constant. This is independent of time and half of the maximum actual velocity of shuttle, that is $2k\omega$.

Nominal acceleration of shuttle $\dfrac{d^2 s}{dt^2} = 0$.

7.7 SHUTTLE CHECKING

The objective of the shuttle checking is to retard the shuttle, nullifying its kinetic energy to zero. Shuttle checking should ensure that the shuttle is stopped within the shuttle box and is not ricocheted back into the shed.

7.7.1 Mechanism of Shuttle Checking

A simple shuttle checking mechanism is shown in Figure 7.13. The incoming shuttle gets rubbed on the spring loaded swell and thereby the frictional force slows down the shuttle. The velocity of the incoming shuttle

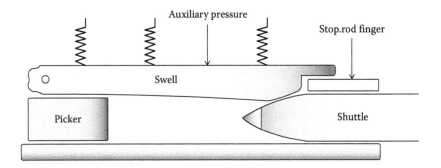

FIGURE 7.13 Checking of shuttle.

is reduced by around 30% by the action of swell. The shuttle is finally stopped as it collides with the picker, which is cushioned by a suitable buffer system.

7.7.2 A Simplified Theoretical Model of Shuttle Checking

Let us assume that the swell is split into two equal parts to avoid the difficulties of asymmetric loading of the shuttle. The forces acting as the shuttle strikes the swell are depicted in Figure 7.14.

Let

F_s is the retarding force acting on the shuttle during its collision with the swell

F_b is the force exerted by the swell on shuttle during the collision

M_s is the mass of the shuttle

M_b is the mass of the swell

e is the coefficient of restitution

μ is the coefficient of friction between the shuttle and the swell

v_s, v_b are the velocities of the shuttle and the swell, respectively, before the collision

v is the velocities of the shuttle and the swell at the time of collision

v'_s, v'_b are the velocities of the shuttle and the swell, respectively, after the collision

The retarding force acting on the shuttle is given by

$$F_s = M_s \frac{dv_s}{dt} \tag{7.17}$$

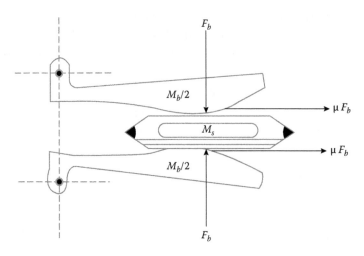

FIGURE 7.14   Forces acting when the shuttle strikes the swell.

The force exerted by each part of the swell on the shuttle during the collision is given by

$$F_b = \frac{1}{2} M_b \frac{dv_b}{dt} \qquad (7.18)$$

At equilibrium

$$F_s = 2\mu F_b$$

or

$$F_s = 2\mu \times \frac{1}{2} M_b \frac{dv_b}{dt}$$

or

$$F_s = \left(\mu M_b \right) \frac{dv_b}{dt} \qquad (7.19)$$

or

$$M_s \frac{dv_s}{dt} = \left(\mu M_b \right) \frac{dv_b}{dt}$$

or

$$M_s dv_s = \left(\mu M_b \right) dv_b \qquad (7.20)$$

Equation 7.20 is similar in form representing the collision of two masses M_s and μM_b which is depicted schematically in Figure 7.15.

Now the coefficient of restitution

$$e = \frac{v'_s - v'_b}{v_s - v_b}$$

$$= \frac{v'_s - v'_b}{v_s - 0} \left(\text{as the velocity of the swell before collision, i.e. } v_b \text{ is zero} \right)$$

or

$$e = \frac{v'_s - v'_b}{v_s}$$

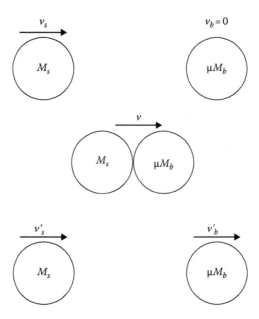

FIGURE 7.15   Schematic representation of shuttle collision with the swell.

or

$$v'_s = v'_b + ev_s \qquad (7.21)$$

From the law of conservation of momentum, $M_s v_s + \mu M_b \times 0 = M_s v'_s + \mu M_b v'_b$

or, $M_s v_s = M_s v'_s + \mu M_b v'_b$.

By substituting the expression of v'_s from Equation 7.21

$$M_s v_s = M_s \left(v'_b + ev_s \right) + \mu M_b v'_b \quad \text{or} \quad M_s v_s = \frac{v'_b \left(M_s + \mu M_b \right)}{1-e} \qquad (7.22)$$

By dividing Equation 7.20 by Equation 7.22,

$$\frac{dv_s}{v_s} = \frac{\mu M_b \left(1-e\right)}{M_s + \mu M_b} \left(\frac{dv_b}{v'_b} \right)$$

The velocity of the swell before collision, that is v_b, is zero.

So

$$\frac{dv_b}{v'_b} = \frac{v'_b - v_b}{v'_b} = 1$$

or

$$\frac{dv_s}{v_s} = \frac{\mu M_b (1-e)}{M_s + \mu M_b}$$ (7.23)

So retardation in velocity of shuttle will be more if the coefficient of friction between the shuttle and the swell is higher and mass of the swell is greater.

7.7.3 Checking by the Action of Picker

Let

F_s is the retarding force acting on the shuttle during its collision with the picker

F_p is the force exerted by the picker against the shuttle during the collision

M_p is the mass of the picker

v_s, v_p are the velocities of the shuttle and the picker, respectively, before the collision

v is the velocities of the shuttle and the picker at the time of collision

v_s', v_p' are the velocities of the shuttle and the picker, respectively, after the collision

The forces acting when the shuttle strikes the picker are shown in Figure 7.16.

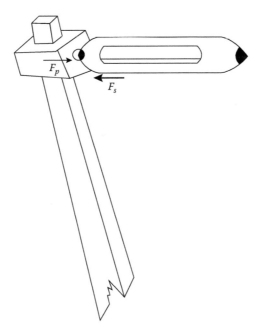

FIGURE 7.16   Forces acting when the shuttle strikes the picker.

The retarding force acting on the shuttle is given by

$$F_s = M_s \frac{dv_s}{dt}$$

The force exerted by the picker against the shuttle during the collision is given by

$$F_p = M_p \frac{dv_p}{dt}$$

At equilibrium

$$F_s = F_p$$

So

$$M_s dv_s = M_p dv_p \qquad (7.24)$$

Equation 7.24 is similar in form representing the collision of two masses M_s and M_p, which is depicted schematically in Figure 7.17.

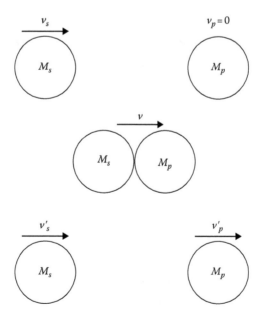

FIGURE 7.17 Schematic representation of shuttle collision with the picker.

Now the coefficient of restitution,

$$e = \frac{v'_s - v'_p}{v_s - v_p} = \frac{v'_s - v'_p}{v_s - 0}$$

(as the velocity of the picker before collision, i.e. v_p, is zero)

or

$$e = \frac{v'_s - v'_p}{v_s}$$

or

$$v'_s = v'_p + ev_s \tag{7.25}$$

From the law of conservation of momentum

$$M_s v_s + M_p \times 0 = M_s v'_s + M_p v'_p$$

or

$$M_s v_s = M_s \left(v'_p + ev_s\right) + M_p v'_p$$

or

$$M_s v_s = \frac{v'_p \left(M_s + M_p\right)}{1-e} \tag{7.26}$$

Equation 7.26 is analogous to Equation 7.22.
By dividing Equation 7.24 by Equation 7.26,

$$\frac{dv_s}{v_s} = \frac{M_p(1-e)}{M_s + M_p}\left(\frac{dv_p}{v'_p}\right)$$

Now

$$\frac{dv_p}{v'_p} = \frac{v'_p - v_p}{v'_p} = \frac{v'_p - 0}{v'_p} = 1$$

or

$$\frac{dv_s}{v_s} = \frac{M_p(1-e)}{M_s + M_p} \tag{7.27}$$

Equation 7.27 is analogous to Equation 7.23. Equations 7.23 and 7.27 imply that the change in shuttle velocity is proportionate to the mass of the checking element, that is swell or picker. In practice, a greater change in shuttle velocity occurs when the shuttle strikes the picker in comparison with that of the swell. For example, around 70% shuttle velocity is reduced by the action of picker, whereas around 30% shuttle velocity is reduced by the action of swell (Lord and Mohamed, 1982).

NUMERICAL PROBLEMS

7.1 Calculate the average shuttle velocity from the following particulars.

> Reed width = 1.5 m
>
> Effective length of the shuttle = 30 cm
>
> Loom speed = 210 picks/min

Degree of crankshaft rotation available for the travel of the shuttle = 140°

Solution:

From the given data

$$R = 1.5 \text{ m}, L = 0.3 \text{ m}, P = 210 \text{ picks/min and } \theta = 140°.$$

$$\text{Shuttle velocity} = v = \frac{6P(R+L)}{\theta} = \frac{6 \times 210 \times (1.5+0.3)}{140} = 16.2 \text{ m/s}$$

So the average shuttle velocity is 16.2 m/s.

7.2 Calculate the loom speed and weft insertion rate of a loom if the average velocity of the shuttle is 50 km/h, the degree of crankshaft rotation available for the travel of shuttle across the shed is 135°, the shuttle length is 40 cm and the reed width is 2 m.

Solution:

From the given data, we get

$$R = 2 \text{ m}, \quad L = 0.4 \text{ m}, \quad \theta = 135°, \quad v = \frac{50 \times 1000}{3600} = 13.9 \text{ m/s}$$

$$\text{Loom speed } P = \frac{v\theta}{6(R+L)} = \frac{13.9 \times 135}{6 \times (2+0.4)} = 130 \text{ picks/min}$$

Weft insertion rate $(\text{WIR}) = P(R+L) = 130(2+0.4) = 312$ m/min

So, loom speed and weft insertion rate are 130 picks/min and 312 m/min, respectively.

7.3 Calculate the power required for picking from the following particulars:

> Reed width = 1.8 m
>
> Effective length of the shuttle = 30 cm
>
> Loom speed = 220 picks/min
>
> Shuttle mass = 500 g

The shuttle enters the shed at 110° and leaves the shed at 240°.

Solution:
From the given data, we get

> $R = 1.8$ m, $L = 0.3$ m, $m = 0.5$ kg, $P = 220$ picks/min,
> $\theta = (240° - 110°) = 130°$

$$\text{Power required for picking} = \frac{3mp^3(R+L)^2}{\theta^2} \times 10^{-4}$$

$$= \frac{3 \times 0.5 \times 220^3(1.8+0.3)^2}{130^2} \times 10^{-4}$$

$$= 0.417 \text{ kW}$$

So the power required for picking is 0.417 kW.

7.4 Calculate the power required for picking and work done per pick from the following parameters:

> Shuttle mass = 0.5 kg
>
> Reed width = 1.5 m
>
> Shuttle length = 0.25 m
>
> Loom speed = 200 picks/min

Degree of crankshaft rotation for the travel of the shuttle = 120

Solution:

Here

$$R + L = (1.5 + 0.25) \text{ m} = 1.75 \text{ m and } m = 500 \text{ g} = 0.5 \text{ kg}.$$

$$\text{Power required for picking} = P = \frac{3mp^3(R+L)^2}{\theta^2} \times 10^{-4} \text{ kW}$$

So

$$P = \frac{3 \times 0.5 \times 200^3 \times (1.75)^2}{120^2} \times 10^{-4} = 0.255 \text{ kW}$$

$$\text{Work done/pick} = \frac{18mP^2(R+L)^2}{\theta^2} \text{ J} = \frac{18 \times 0.5 \times 200^2 \times (1.75)^2}{120^2} = 76.56 \text{ J}$$

So power required for picking is 0.255 kW and work done per pick is 76.56 J.

7.5 A shuttle of 500 g mass is uniformly accelerated by the picker of a cone under-pick system. The length of the picker stroke is 20 cm. The final velocity of the shuttle is 15 m/s. If AB: BC = 1:2 in Figure 7.18, then calculate the force acting on the picking strap (lug strap) during picking.

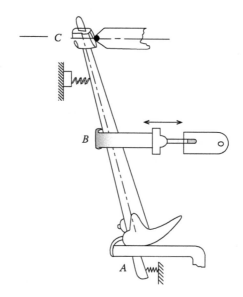

FIGURE 7.18   Cone under-pick mechanism.

Solution:

For uniform acceleration

$$v^2 = 2fs$$

where
    v is the final velocity of the shuttle
    f is the acceleration of shuttle
    s is the distance travelled by the shuttle during acceleration

$$f = \frac{v^2}{2s} = \frac{15^2}{2 \times 0.2} = 562.5 \text{ m/s}^2$$

Force applied by the picker on shuttle $= P = mf$

$$= \frac{1}{2} \times 562.5 = 281.25 \text{ N}$$

Let $AB = x$ and $BC = 2x$, So $AC = 3x$.
    If F is the force acting on the lug-strap, then

$$P \times AC = F \times AB$$

or

$$P \times 3x = F \times x$$

So

$$F = 3P = 3 \times 281.25 = 843.75 \text{ N}.$$

So

The force acting on the lug-strap is 843.75 N.

7.6 Calculate the alacrity of picking system, the maximum acceleration of shuttle and effective stroke of picker considering straight-line
nominal displacement and the following data.

Loom speed = 180 picks/min

Velocity of the shuttle when it leaves the picker = 15 m/s

Degree of crankshaft rotation available for picking mechanism = 30°

Shuttle mass = 0.5 kg

Plot the nominal and actual displacement, velocity and acceleration
against time.

Solution:

From the given data

$$\omega = \frac{180 \times 2\pi}{60} = 18.86 \text{ rad/s}$$

and

$$\text{Time taken by picking mechanism} = \frac{60}{180} \times \frac{30}{360} = 0.0278 \text{ s.}$$

The shuttle attains its maximum actual velocity when it leaves the picker.

So

$$\left(\frac{dx}{dt}\right)_{max} = 15 \text{ m/s.}$$

Shuttle velocity becomes maximum at $t = \pi/n$

So

$$n = \frac{\pi}{0.0278} = 113.05 \text{ s}^{-1}$$

or

$$\sqrt{\frac{\eta}{m}} = 113.05 \text{ s}^{-1}$$

This gives the value of $\eta = 6390.2$ N/m $= 6.39$ kN/m (as shuttle mass $m = 0.5$ kg).

Now

$$\left(\frac{dx}{dt}\right)_{max} = 15$$

or

$$2k\omega = 15$$

or

$$k = \frac{15}{2 \times 18.86} = 0.3976$$

or

$$\left(\frac{d^2x}{dt^2}\right)_{max} = k\omega n = 0.3976\times18.86\times113.05 = 847.7 \text{ m/s}^2$$

From Equation 7.13, we get

$$L = \frac{\pi k\omega}{n}$$

or

$$L = \frac{\pi\times0.3976\times18.86}{113.05} = 0.208 \text{ m} = 20.8 \text{ cm}.$$

So the alacrity of picking system, the maximum acceleration of shuttle and the effective stroke of picker are 113.05 s^{-1}, 847.7 m/s^2 and 20.8 cm, respectively.

The plots of nominal and actual displacement, velocity and acceleration are depicted in Figures 7.19 through 7.21 (for straight-line nominal movement).

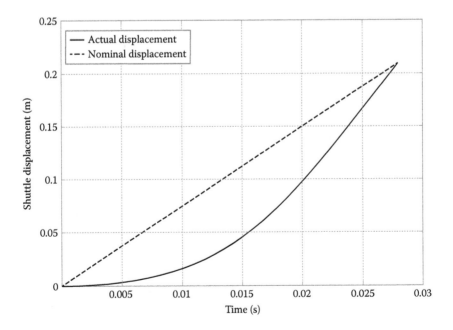

FIGURE 7.19 Nominal and actual displacement of the shuttle.

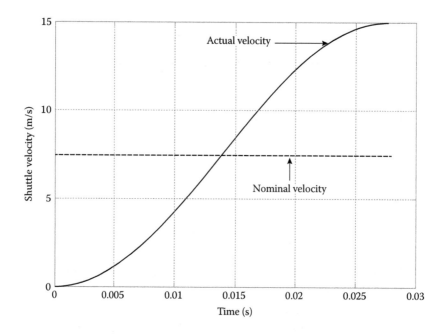

FIGURE 7.20   Nominal and actual velocity of the shuttle.

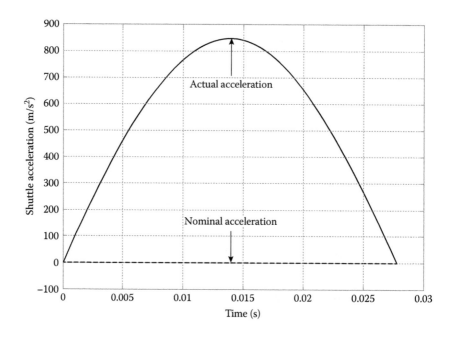

FIGURE 7.21   Nominal and actual acceleration of the shuttle.

REFERENCES

Banerjee, P. K. 2015. *Principles of Fabric Formation*. Boca Raton, FL: CRC Press.

Lord, P. R. and Mohamed, M. H. 1982. *Weaving: Conversion of Yarn to Fabric*, 2nd edn. Merrow, UK: Merrow Technical Library.

Marks, R. and Robinson, A. T. C. 1976. *Principles of Weaving*. Manchester, UK: The Textile Institute.

CHAPTER 8

Picking in Shuttleless Looms

8.1 LIMITATIONS OF SHUTTLE LOOM

- The mass of a shuttle with pirn is around 500 g. The mass of yarn required for one pick generally lies within 0.01–0.2 g. Therefore, accelerating and decelerating a mass which is 2,500–50,000 times higher than that of a pick to be inserted consumes huge energy. Besides, the energy required for picking is proportional to the product of shuttle mass and third power of loom speed (Equation 7.7). Therefore, power consumption in shuttle loom increases almost exponentially if loom speed is increased. This limits the maximum possible speed of shuttle loom.

- The height of the shuttle is around 30–35 mm. Therefore the sley must recede sufficiently so that enough space is created between the top and bottom shed lines for the easy entry of the shuttle. If it is assumed that the average velocity of shuttle during picking is 15 m/s and loom width is 1.5 m, then the shuttle will take 0.1 s time to travel through the shed. The shuttle remains within the shed from 110° to 240° angular position of the crank, causing the shuttle flight duration of 130° which is approximately one-third of a complete pick cycle (360°). Therefore, the duration of a pick cycle will be around 0.28 s, so the loom speed will be around 217 rpm (picks per minute). The shuttle cannot enter the shed before 110° as the shed is not open enough for its entry. The shuttle cannot leave the shed much later than 240° due to the constraint imposed by the fast reed motion. Thus the degree of crankshaft rotation (θ) available for the travel of shuttle cannot be increased, which is one of the ways to increase the loom speed keeping the velocity of shuttle constant (Equation 7.4).

- As the weaving continues, mass of the pirn reduces. Therefore, velocity of the shuttle also changes periodically, causing some variation in weft tension as the latter is dependent on the unwinding speed of yarn from the pirn.

- The yarn content in a pirn is very limited (around 50 g). Therefore, automatic looms are equipped with the pirn changing mechanism. Frequent changes of pirn, which may have been wound from different cones, can create quality problems in fabric. If the count or hairiness variation between two consecutive pirns is very high, then the produced fabric may have weft bars.

- The noise level of a shuttle loom is around 100–110 dB and thus the noise level inside a loom shed is almost deafening. This is ergonomic limitation of shuttle loom.

Shuttleless looms overcome most of the aforesaid limitations. The technologies of weft insertion in shuttleless looms, which have seen commercial success, can be classified under four heads (Banerjee, 2015):

1. Projectile (partially guided solid carrier)
2. Rapier (fully guided solid carrier)
3. Air-jet (guided fluid carrier)
4. Water-jet (unguided fluid carrier)

Small weft carriers are used in projectile looms for the insertion of pick. The mass of a projectile is around 40–50 g. Rapier looms use flexible or rigid carriers which transport the pick across the shed. Fluid looms (air-jet and water-jet) utilise fluid drag created either by air or by water for the insertion of pick. In contrast to shuttle loom, all the weft carriers in shuttleless looms carry the weft yarn which is sufficient for one pick. Besides, weft insertion is done from one side of the loom (generally from the left-hand side). Air-jet is the most popular, whereas rapier is the most versatile among the shuttleless weaving systems. Air-jet loom consumes the maximum power to produce one square meter of fabric, followed by rapier, shuttle, projectile and water-jet.

8.2 PROJECTILE PICKING SYSTEM

It has already been mentioned that the mass of a projectile is around 40–50 g, whereas it is around 500 g for a conventional shuttle. The projectile, also known as gripper, carries the weft yarn sufficient for one pick. The projectile picking system is based on storing the strain energy by subjecting a metallic rod to torsional deformation and then releasing the energy at an opportune moment to accelerate the weft carrier. If a rod is twisted, the energy that will be stored inside the rod will be equal to the work done during twisting. Now, if the rod is allowed to recover from the deformation, the stored energy can be utilised to accelerate the projectile. Therefore, it is important to understand the deformation of a rod under torsion.

8.2.1 Torsion of a Circular Rod

If one end of a circular rod is firmly fixed and some torque is applied at the free end, then the rod will undergo a torsional deformation about its own axis. The rod, under this circumstance, has some shearing strain due to torsion. The shape of the circular rod changes but its volume remains the same. As a result of the shearing strain, a restoring couple is developed inside the material of the rod due to the elastic property of the material. In the equilibrium condition, the moment of the torsional couple is equal and opposite to the moment of the restoring couple.

Consider a circular rod of length l and radius r as shown in Figure 8.1. The upper end of the circular rod is fixed. The twisting torque is applied at the lower end of the rod. The radius OP on the lower surface of the rod has moved through an angle θ, which is known as the angle of twist. So POP′ = θ. Now the moment of the restoring couple developed inside the material of the rod by virtue of its elastic property will be calculated.

Consider a line CP on the circumference of a thin cylindrical shell of radius x and thickness dx, parallel to the axis of the circular rod (OO′). The line CP moves to CP′ during twisting, turning an angle φ. In this case, φ is the angle of shear. The displacement PP′ will be the maximum on the circumference of the circular rod and it will be less towards the centre O.

Now the solid circular rod may be assumed to be made of a large number of such coaxial cylindrical shells. All the radii of lower end of the circular rod will turn through equal angle θ. However, the displacement PP′ will be maximum on the circumference of the circular rod and will be gradually

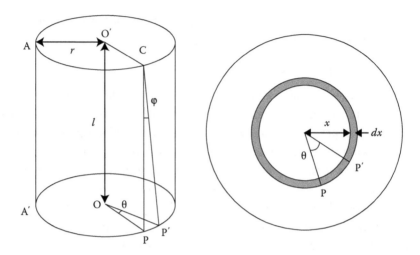

FIGURE 8.1   Torsion in a circular rod.

less towards the centre, becoming zero exactly at the centre O. Therefore, the shear angle φ will have the maximum value when $x = r$ and the minimum value when $x = 0$. It implies that the shearing strain is not equal everywhere.

If P and P′ are supposed to be situated on the circumference of the cylindrical shell of radius x, then following expression can be written from Figure 8.1.

$$PP' = l\varphi = x\theta$$

If G is the modulus of rigidity of the material of the rod, then

$$G = \frac{\text{Shear stress}}{\text{Shear strain}} = \frac{F}{\varphi},$$

where F is the shear stress.

So

$$F = G\varphi = \frac{Gx\theta}{l}$$

Now, the area of the circular ring of radius x and thickness $dx = 2\pi x\,dx$

So the total force on this area = stress × area = $\dfrac{Gx\theta}{l} \times 2\pi x\,dx = \dfrac{2\pi G\theta}{l} \times x^2\,dx$

The moment of this force about the axis OO′ of the circular rod $= \dfrac{2\pi G\theta}{l}$

$$\times x^2 dx \times x = \dfrac{2\pi G\theta}{l} \times x^3\ dx$$

Total moment of the torsional couple

$$T = \int_0^r \dfrac{2\pi G\theta}{l} x^3\ dx = \dfrac{2\pi G\theta}{l}\left[\dfrac{x^4}{4}\right]_0^r = \dfrac{\pi G\theta r^4}{2l}$$

Therefore

$$T = \dfrac{\pi G\theta r^4}{2l} \tag{8.1}$$

The moment of the torsional couple for unit twist in radian is $\pi G r^4/2l$. This is known as torsional rigidity (τ) of the material of the circular rod (N·m/rad). So

$$\tau = \dfrac{\pi G r^4}{2l} \tag{8.2}$$

If the circular rod is hollow and has r_1 and r_2 as internal and external radii, respectively, then the moment of the torsional couple will be as follows:

$$T_{hollow} = \int_{r_1}^{r_2} \dfrac{2\pi G\theta}{l} x^3 dx = \dfrac{\pi G\theta}{2l}\left(r_2^4 - r_1^4\right) \tag{8.3}$$

8.2.2 Principle of Torsion Rod Picking Mechanism

Figure 8.2 shows the schematic view of torsion rod picking mechanism used in projectile loom.

The picking system can be considered as a three-bar link mechanism composed of toggle lever, link and torsion lever. The positions of fulcrums for toggle lever and torsion lever are fixed. The picking shaft completes one rotation in every pick and thus the nose of the picking cam presses the anti-frictional roller, which is attached with the toggle lever, towards the right-hand side. This causes the clockwise movement of the toggle lever

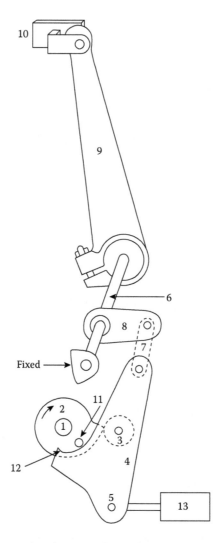

FIGURE 8.2 Torsion rod picking mechanism. 1: Picking shaft, 2: picking cam, 3: anti-friction roller, 4: toggle lever, 5: fulcrum point for toggle lever, 6: torsion rod, 7: link, 8: torsion lever, 9: picking lever, 10: picker, 11. small roller on picking cam, 12: contour of toggle lever, 13: oil brake system. (From Marks, R. and Robinson, A.T., *Principles of Weaving*, The Textile Institute, Manchester, UK, 1976.)

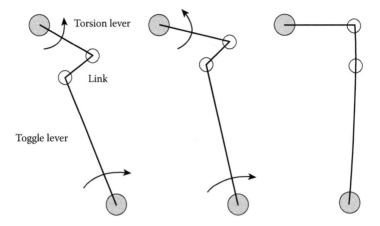

FIGURE 8.3 Schematic representation of twisting of torsion rod.

about its fulcrum point. Thus the torsion lever gets anti-clockwise rotation through the link. Therefore, the free end of the torsion rod is twisted. When the picking cam tip loses contact with the anti-frictional roller, the twist in the torsion rod becomes the maximum. However, the link system does not collapse as the axis of link, through which the torsion rod and torsion lever apply clockwise torque, passes through the right-hand side of fulcrum of toggle lever. The system collapses when the small roller on picking cam passes over the upward contour of toggle lever and thus rotates the toggle lever in anti-clockwise direction. As soon as the axis of the link passes through the left-hand side of fulcrum of toggle lever, the link system collapses instantly and the torsion rod rotates clockwise releasing all the stored energy. Thus the picking lever rotates clockwise and the picker hits the projectile. The residual energy of the system, which is around 60% of the stored energy, is absorbed by the oil brake system.

Figure 8.3 shows the schematic representation of movements of toggle lever, link and torsion lever during the twisting of torsion rod. As the toggle lever rotates clockwise, the link becomes more upright and the torsion lever rotates anti-clockwise. The situation becomes exactly opposite when the torsion rod is untwisted.

8.2.3 Acceleration and Deceleration of Picker and Projectile

Figure 8.4 shows the acceleration and deceleration profile of picker during the picking in an older version of projectile loom. The acceleration and deceleration are far from being uniform. The picker shoe remains in acceleration mode for a displacement of 65 mm. The projectile attains its

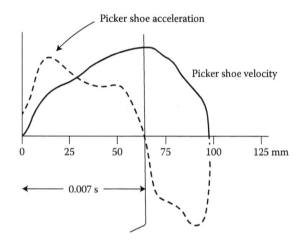

FIGURE 8.4   Velocity and acceleration profile of picker shoe. (From Marks, R. and Robinson, A.T., *Principles of Weaving*, The Textile Institute, Manchester, UK, 1976.)

peak velocity of 24 m/s at this point and then loses contact with the picker. The picker is decelerated over a distance of around 33 mm. The peak acceleration of the picker is around 6600 m/s². For uniform acceleration, the value will be around 4400 m/s² considering peak velocity and corresponding displacement of 24 m/s and 65 mm, respectively. Peak deceleration is around 9900 m/s², which is also much higher than that obtained with uniform deceleration.

Figure 8.5 shows the velocity profile of a projectile (Marks and Robinson, 1976). It acquires the velocity of 24 m/s at 0.007 s. It travels

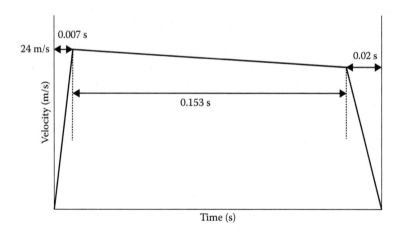

FIGURE 8.5   Velocity profile of projectile.

across the shed in 0.153 s. The time for deceleration is 0.02 s. Neglecting the retardation of projectile during its movement through the shed, total distance covered by the projectile in one pick is as follows:

$$\text{Distance} = \left[\frac{1}{2} \times 0.007 \times 24 + 24 \times 0.153 + \frac{1}{2} \times 0.02 \times 24 \right] \text{m} = 3.996 \text{ m}$$

If the time available for projectile movement across the shed is equivalent to 150° rotation of loom shaft, then the time for one pick cycle is = (360/150) × 0.153 s = 0.367 s.

So

$$\text{The picks per minute or loom rpm} = \frac{60}{0.367} \approx 164.$$

Reed width = Average velocity of projectile × time required for the projectile to travel across the reed = 24 × 0.153 = 3.672 m.

$$\text{Weft insertion rate} = \text{Reed width} \times \text{picks per minute}$$

$$= 3.672 \times 164 = 602.2 \text{ m/min}.$$

8.2.4 Hypothetical Velocity Profile of Picker

Figure 8.6 shows two hypothetical situations of picker movement resulting in the same final velocity of the picker and projectile. In the first case, the picker moves with uniform acceleration during picking and thus it follows the Newton's laws of motion. In the second case, the acceleration of picker is time-dependent. The acceleration increases and decreases linearly with time in the first half (from $t = 0$ to $t = T/2$) and the second half (from $t = T/2$ to $t = T$) of picking, respectively. The slope of the line representing acceleration is u and $-u$, in the first half and second half, respectively.

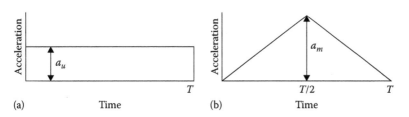

FIGURE 8.6   (a) Uniform and (b) non-uniform acceleration of picker.

8.2.4.1 Uniform Acceleration

In the case of uniform acceleration, the acceleration is a_u and total time of picking is T.

So

$$\text{Velocity } V = \int a_u \, dt = a_u t + C_1$$

where C_1 is a constant.

At time $t = 0$, velocity $V = 0$. Therefore, $C_1 = 0$.

Hence

$$V = a_u t$$

$$\text{Displacement } S = \int \left(a_u t \right) dt = \frac{a_u t^2}{2} + C_2$$

At time $t = 0$, displacement $S = 0$. Therefore, $C_2 = 0$.

So

$$S = \frac{a_u t^2}{2}$$

So the total displacement at the end of picking, when time $t = T$, is $a_u T^2/2$.

For both the cases, the area under the acceleration–time curve will be same as the final velocity of the picker is same.

$$\text{For uniform acceleration: Area} = a_u \times T$$

$$\text{For non-uniform acceleration: Area} = \frac{1}{2} \times T \times a_m$$

where a_m is peak acceleration in the case of non-uniform acceleration.

So

$$a_u \times T = \frac{1}{2} \times T \times a_m$$

or

$$a_u = \frac{a_m}{2}$$

Therefore

$$\text{Total displacement of the picker} = S(\text{total}) = \frac{a_m}{2} \times \frac{T^2}{2} = \frac{a_m T^2}{4}$$

8.2.4.2 Non-Uniform Acceleration

In the case of non-uniform acceleration, the expression for acceleration is as follows.

$$\text{Acceleration } a = \begin{cases} mt & \text{for } 0 \le t \le \dfrac{T}{2} \\[3mm] a_m - m\left(t - \dfrac{T}{2}\right) & \text{for } \dfrac{T}{2} \le t \le T \end{cases}$$

where m is the slope of line representing non-uniform acceleration.

Now

$$a_m = \frac{mT}{2}$$

So

$$a = \begin{cases} mt & \text{for } 0 \le t \le \dfrac{T}{2} \\[3mm] 2a_m - mt & \text{for } \dfrac{T}{2} \le t \le T \end{cases}$$

First half of linear acceleration between time $t = 0$ and time $t = T/2$

$$V_1 = \int a \, dt = \int mt \, dt = \frac{mt^2}{2} + C_3$$

At time $t = 0$, velocity $V_1 = 0$. Therefore, $C_3 = 0$.

Therefore

$$V_1 = \frac{mt^2}{2}$$

So

$$\text{Velocity at time } t = T/2 = V_{1(T/2)} = \left[\frac{mt^2}{2}\right]_0^{\frac{T}{2}} = \frac{mT^2}{8} = \frac{a_m T}{4}$$

$$\text{Displacement } S_1 = \int V_1 \, dt = \int \frac{mt^2}{2} \, dt = \frac{mt^3}{6} + C_4$$

At time $t = 0$, displacement $S_1 = 0$. Therefore, $C_4 = 0$.

Hence, the total displacement at the end of the first half, when $t = T/2$, is

$$S_1(\text{total}) = \left[\frac{mt^3}{6}\right]_0^{\frac{T}{2}} = \frac{mT^3}{48}$$

Since

$$a_m = \frac{mT}{2}$$

therefore

$$S_1(\text{total}) = \frac{a_m T^2}{24}$$

For the second half of movement with linear deceleration of picker,

$$V_2 = \int (2a_m - mt) \, dt$$

or

$$V_2 = 2a_m t - \frac{mt^2}{2} + C_5$$

At time $t = T/2$, velocity $V_2 =$ velocity V_1.

So

$$\frac{a_m T}{4} = \frac{2a_m T}{2} - \frac{mT^2}{8} + C_5$$

or

$$C_5 = -\frac{a_m T}{2}$$

So

$$V_2 = 2a_m t - \frac{mt^2}{2} - \frac{a_m T}{2} = 2a_m t - \frac{2a_m}{T} \times \frac{t^2}{2} - \frac{a_m T}{2} = 2a_m t - \frac{a_m t^2}{T} - \frac{a_m T}{2}$$

or

$$S_2 = \int \left(2a_m t - \frac{a_m t^2}{T} - \frac{a_m T}{2} \right) dt$$

$$S_2 = a_m t^2 - \frac{a_m t^3}{3T} - \frac{a_m T t}{2} + C_6$$

At time $t = T/2$, displacement $S_2 =$ displacement $S_1 = a_m T^2/24$
    Hence

$$\frac{a_m T^2}{24} = \frac{a_m T^2}{4} - \frac{a_m}{3T} \times \frac{T^3}{8} - \frac{a_m}{2} \times T \times \frac{T}{2} + C_6$$

or

$$C_6 = \frac{a_m T^2}{12}$$

Therefore

$$S_2 = a_m t^2 - \frac{a_m t^3}{3T} - \frac{a_m T t}{2} + \frac{a_m T^2}{12}$$

So the total displacement between time $T/2$ and T will be

$$S_2(\text{total}) = a_m T^2 - \frac{a_m T^2}{3} - \frac{a_m T^2}{2} + \frac{a_m T^2}{12} - \frac{a_m T^2}{4} + \frac{a_m T^2}{24}$$

$$+ \frac{a_m T^2}{4} - \frac{a_m T^2}{12} = \frac{5}{24} a_m T^2$$

From time $t = 0$ to $t = T$, the total displacement of the picker = S_1(total) + S_2(total)

$$= \frac{a_m T^2}{24} + \frac{5a_m T^2}{24} = \frac{a_m T^2}{4} = S(\text{total})$$

Therefore, the displacements are equal in both the cases.

8.2.5 Sequence of Weft Insertion in Projectile Loom

The sequence of weft insertion in projectile loom is schematically shown in Figures 8.7 through 8.13 (Adanur, 2001).

The arrangement of essential elements from the left to right is as follows:

- Weft package
- Weft accumulator
- Weft brake
- Tension lever
- Projectile feeder
- Picker
- Weft cutter
- Weft gripper

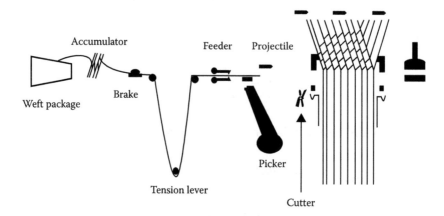

FIGURE 8.7   First step of weft insertion.

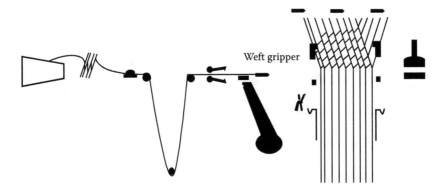

FIGURE 8.8   Second step of weft insertion.

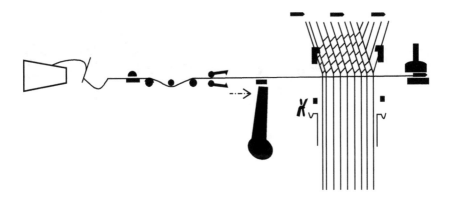

FIGURE 8.9   Third step of weft insertion.

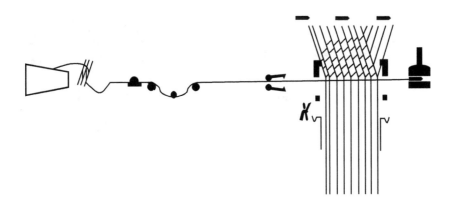

FIGURE 8.10   Fourth step of weft insertion.

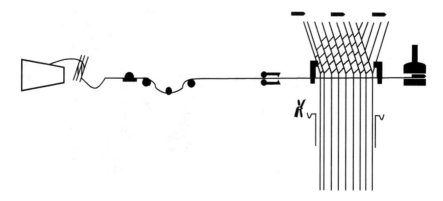

FIGURE 8.11   Fifth step of weft insertion.

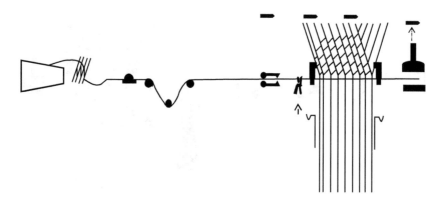

FIGURE 8.12   Sixth step of weft insertion.

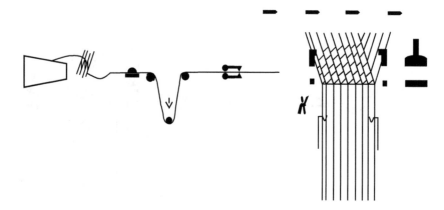

FIGURE 8.13   Seventh step of weft insertion.

In Figure 8.7, the projectile feeder presents the yarn tip to the projectile which has taken up the position for picking. The weft brake is closed. Tension lever accommodates some extra yarn length indicated by the elongated 'U'-shape.

In Figure 8.8, the projectile feeder has opened as the tip of the yarn has been firmly gripped by the projectile. The brake is about to open and the system is ready for picking.

In Figure 8.9, the brake is open and the picking system has propelled the projectile across the shed. The yarn coils are released from the accumulator. The tension lever also facilitates the weft insertion by changing its position and thus converting the yarn loop into straight segment. When the pick crosses a certain width of the shed, the brake is partially lowered to create some tension on the pick and prevent overfeeding of yarn.

The projectile is stopped at the receiving side and then it recedes a bit as shown in Figure 8.10. The tension lever adjusts the tension by accommodating the extra length of yarn. This is indicated by the small loop of yarn around the tension lever. The weft feeder, in open condition, moves towards the right-hand side, that is towards the selvedge.

The projectile feeder closes and grips the pick as shown in Figure 8.11. The pick is also gripped by the two grippers close to the left and right selvedges of the fabric.

The projectile releases the pick at the receiving side (as shown in Figure 8.12) and then drops on the returning conveyor. The cutter cuts the pick near the left-hand side selvedge. So the pick is now ready to be beaten up to the cloth fell.

The beat-up is performed as shown in Figure 8.13. The projectile feeder returns to the left-hand side in closed condition holding the tip of the pick. The tension lever accommodates the extra length of yarn indicated by a relatively bigger loop. The new projectile takes the position for the next picking cycle.

8.2.6 Loom Timing

The sley dwell in projectile loom spans over more than half of the pick cycle. The extent of dwell varies with the loom width with higher width requiring higher dwell period and vice versa. The duration of sley dwell can vary from 190° to 230° rotation of loom shaft. For small width loom (190 cm), the duration of sley dwell has a span of around 190°, whereas for wide width loom (350 cm and higher), it is around 230° (Figure 8.14).

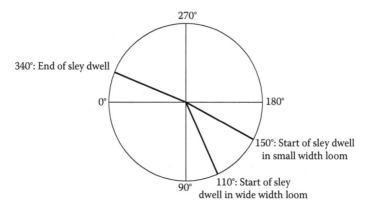

FIGURE 8.14 Sley dwell in projectile loom.

8.2.7 Beat-Up Mechanism

As opposed to shuttle loom where beat-up is done by a crank system, a cam-driven system is used in projectile loom. In a crank-driven system, the sley moves continuously in the entire pick cycle. However, the movement profile deviates from that of simple harmonic motion as the sley stays near the back centre of the loom for greater duration to facilitate easy travel of the shuttle. In contrast, a cam-driven sley remains absolutely idle during the dwell period and reciprocates only during the remaining part of the pick cycle. The dwell of sley facilitates the travel of the projectile through the shed as the projectile guide teeth or rakes mounted on the sley also remains stationary. Sley dwell becomes possible because the sweep of the sley, which is determined by the dimension of the weft carrier, is much lower in projectile loom as compared to that of shuttle loom. Besides, the mass of the sley in projectile loom is much lower than that of shuttle loom. Therefore, the lack of available time for acceleration and lower mass compensates each other while determining the force required for acceleration and deceleration of the sley assembly.

The obvious question that arises here is the possibility of using a cam-driven sley in shuttle loom. The probable reasons which preclude that possibility are as follows:

- The sweep of the sley has to be relatively large to create enough space for the unobstructed shuttle movement.

- The dwell of the cam-driven sley will leave relatively small time for the sley to cover the necessary sweep, causing very high acceleration and deceleration.

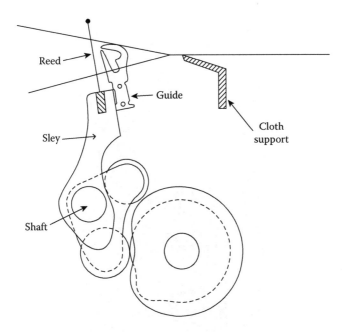

Reed

Guide

Cloth support

Sley

Shaft

FIGURE 8.15 Cam beat-up system. (From Marks, R. and Robinson, A.T., *Principles of Weaving*, The Textile Institute, Manchester, UK, 1976.)

- The higher mass of the sley in shuttle loom will further complicate the situation by increasing the forces required for the high acceleration and deceleration.

Figure 8.15 shows the cam beat-up system. The 'L'-shaped lever carries two rollers or followers. These rollers remain in contact with their respective cams which have significant dwell period. The L-lever is attached with a shaft which also carries the sley. The reed and projectile guide teeth (Figure 8.16) are attached to the sley. If it is assumed that the cams are rotating clockwise, then the lower roller will be pushed towards the left-hand side by the cam shown with broken lines. The depression on the surface of other cam will accommodate the movement of the upper roller towards the right-hand side. As a result, the L-lever and sley moves clockwise. Therefore, beat-up is performed by the reed as the projectile guide teeth move below the bottom shed line.

The differences between crank and cam beat-up can be summarised as given in Table 8.1.

FIGURE 8.16   Projectile guide teeth.

TABLE 8.1   Differences between Crank and Cam Beat-Up

Crank Beat-Up	Cam Beat-Up
The sley moves continuously in the entire pick cycle.	The sley remains stationary during the dwell period.
Time available for the acceleration and deceleration of sley is more.	Time available for the acceleration and deceleration is less as sley dwell occupies more than half of the pick cycle.
Preferred when sweep of the sley is more.	Preferred when sweep of the sley is less.
Used in shuttle loom.	Used in many shuttleless looms, including projectile loom.

8.3 AIR-JET PICKING SYSTEM

In air-jet weaving, the pick is inserted by blowing compressed air (6–7 bar) through nozzles and creating air drag on the weft yarn. The evolution of air-jet weaving technology can be summarised as follows (Vangheluwe, 1999).

- 1914: First U.S. patent was filed by Brooks.

- 1929: New patent was filed by Ballou with suction nozzle in the receiving side.

TABLE 8.2   Number of Air-Jet Loom
Manufacturers

Year	Number of Manufacturers
1975	2
1979	5
1983	11
1987	12
1991	15
1995	9

- 1945: First commercial Maxbo loom was developed by Max Paabo in Sweden. The width of the first loom was 28 cm. His second and third machines were 70 and 120 cm, respectively, in width.

- 1945: Twenty air-jet looms were manufactured in erstwhile Czechoslovakia with confusor.

- 1967: Relay nozzle was presented by Strake at ITMA.

- 1991: Profiled reed and relay nozzle became standard feature of air-jet loom and confusor systems were abandoned.

With the success of air-jet weaving technology after the 1970s, more and more machine manufacturers ventured in the production of this machine. However, consolidation took place and the number of air-jet loom manufacturers dwindled after 1991. Table 8.2 presents the number of air-jet loom manufacturers in different years (Vangheluwe, 1999).

Currently Picanol (Belgium), Dornier (Germany), Toyota (Japan), Tsudakoma (Japan) and Somet (Italy) are some of the major manufacturers of air-jet loom. This list is non-exhaustive.

8.3.1 Bernoulli's Theorem

French mathematician Bernoulli established a theorem in 1738 to relate the velocity and pressure of a fluid in streamline flow. The theorem states that in a steady flow, the sum of all forms of energy (kinetic, potential and internal) in a fluid along a streamline remains constant at all locations. Therefore, the increase in velocity of the fluid occurs simultaneously with a decrease in pressure or a decrease in the potential energy of the fluid.

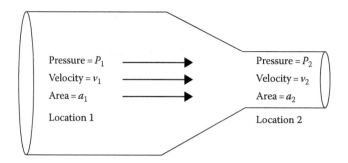

FIGURE 8.17   Fluid flow through a Venturi tube.

The following expression summarises Bernoulli's theorem:

$$\frac{\rho v_1^2}{2} + h_1 \rho g + P_1 = \frac{\rho v_2^2}{2} + h_2 \rho g + P_2 \tag{8.4}$$

where
v_1 and v_2 are the velocity of the fluid at locations 1 and 2, respectively
(Figure 8.17)
P_1 and P_2 are the pressure at locations 1 and 2, respectively
h_1 and h_2 are the height or elevation of locations 1 and 2, respectively,
above the reference plane
g is the acceleration due to gravity
ρ is the density of the fluid

For steady flow, the mass of the fluid flowing through any section of the
Venturi tube per unit time is constant. Considering the fluid to be incom-
pressible, the volumes of fluid flowing per unit time through the sections
having areas of α_1 and α_2 should be equal.
Therefore

$$\alpha_1 v_1 = \alpha_2 v_2 \tag{8.5}$$

As α_1 is greater than α_2, v_2 will be greater than v_1. Therefore, velocity of
the fluid at location 2 will be higher as compared to that at location 1.
However, this gain in fluid velocity will be at the cost of pressure drop at
location 2.

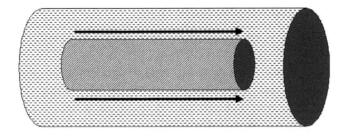

FIGURE 8.18   Fluid drag on a cylindrical yarn segment.

8.3.2 Fluid Drag

Any object inside a fluid will experience a drag if the velocity of the object is different from that of fluid. A swimmer swimming in steady water will experience drag against his or her movement. On the other hand, a swimmer just floating on a moving stream will experience drag in the direction of moving stream. Figure 8.18 shows a cylindrical yarn segment inside a tube filled with fluid. The fluid inside the tube is moving in the direction shown in the figure. The yarn segment will experience a drag depending on the relative velocity of the fluid and yarn segment.

The longitudinal drag experienced by the cylindrical yarn segment can be expressed by the following equation:

$$\text{Drag force} = \frac{1}{2}C_d\rho A\left(v_f - v_y\right)^2 \tag{8.6}$$

where
   C_d is drag coefficient
   ρ is the density of the fluid
   A is the active surface area of the yarn
   v_f and v_y are the velocity of fluid and yarn, respectively

Here

$$A = \pi dl \tag{8.7}$$

where
   d is yarn diameter
   l is the length of the yarn on which drag is acting

8.3.3 Velocity and Acceleration of Pick

It can be inferred from Equation 8.5 that when the difference between the velocities of fluid and yarn is more, that is $v_f \gg v_y$, the drag acting on the yarn is higher. This drag will tend to accelerate the yarn and as a result the relative velocity between the fluid and yarn will reduce. When the velocity of the fluid and yarn will be the same, no drag will be acting on the yarn. Figure 8.19 depicts the change in air and pick (weft) velocity across the reed in a loom which does not have any supplementary device to maintain the air steam. The air velocity reduces exponentially as the distance increases from the air-jet nozzle. Before the point P, the velocity of air is higher than that of pick and thus drag is created which is responsible for the acceleration of the pick. After the critical point P, where the velocity of air and pick is the same, the velocity of the air becomes lower than that of the pick. So the pick tries to maintain its attained velocity due to inertia but loses a bit due to drag created in the direction opposite to its movement. Therefore, the velocity of the tip of pick becomes lower than that of the tail of pick. This may lead to buckling of pick, which is not desirable. Therefore, with a single jet and without any supplementary device to maintain the air stream velocity, it becomes difficult to insert picks in looms having pragmatic width. Numerous attempts have been made to maintain the air stream velocity across the loom.

Luenenschloss and Wahhoud (1984) ascertained the position of the pick during weft insertion in air-jet loom using optical sensors. The acceleration, velocity and time have been shown along the Y-axis with solid, broken and dotted lines, respectively, as shown in Figure 8.20. The pick is subjected to very high acceleration initially which causes a rapid rise in velocity.

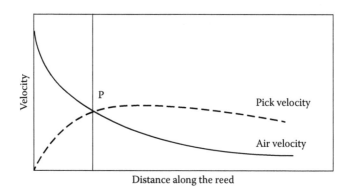

FIGURE 8.19   Velocity of air and pick.

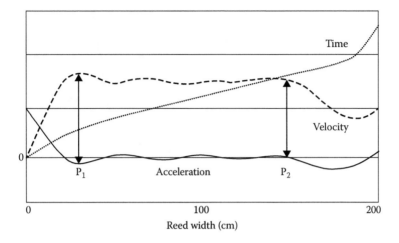

FIGURE 8.20   Velocity and acceleration profiles of pick in air-jet loom.

The acceleration continues till the point P_1 and thus the highest pick velocity is achieved at this point. Between points P_1 and P_2, the pick moves in steady state which implies that the acceleration is almost zero and velocity is more or less constant. The weft brake is applied at point P_2 which leads to drastic deceleration (negative acceleration) in the pick. As a result, the velocity also drops sharply after point P_2. Towards the end, the weft velocity rises again due to the action of the stretch or suction nozzle.

8.3.4 Devices to Control the Air Flow

8.3.4.1 Guide Plates

Guides plates were used in Maxbo loom to maintain the velocity of air stream. In this case, two metallic guide plates along with the reed form a tunnel having triangular cross section as shown in Figure 8.21. The design of the system is cumbersome and it causes problem in beat-up.

8.3.4.2 Confusor

Figures 8.22 and 8.23 show the confusors which are like metallic annular rings with a cut at the upper side. The confusors are positioned one after another in a dense array to create a tunnel which helps maintain the velocity of air stream over a greater distance. As confusors are attached with the sley, the former comes out of the shed through the bottom shed line during the beat-up, leaving behind the newly inserted pick within the shed. Therefore, confusors create lot of abrasion with the warp yarns in every pick. The problem becomes more in the case of higher warp density.

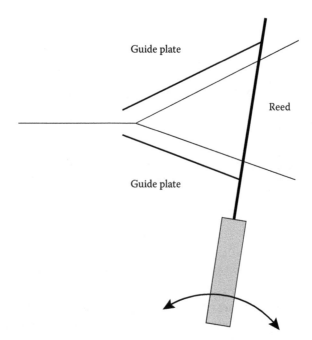

FIGURE 8.21   Guide plates.

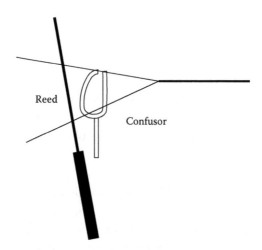

FIGURE 8.22   Confusor and reed. (From Vangheluwe, L., *Air-Jet Weft Insertion (Textile Progress)*, The Textile Institute, Manchester, UK, 1999.)

FIGURE 8.23   Confusor.

8.3.4.3 Profile Reed

All the present variants of air-jet loom use relay nozzles and profile reed as the combination has become a standard technology to maintain the velocity of air stream. The profile reed can be considered as the amalgamation of confusor and reed. In the case of profile reed, the reed dents have been so designed that when they are arranged in a line, they form a tunnel as shown in Figure 8.24. Thus profile reed is also known as tunnel reed.

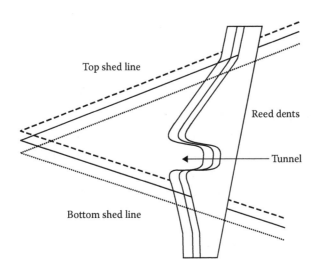

FIGURE 8.24   Profile reed.

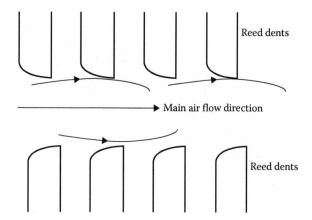

FIGURE 8.25 Front view of profile reed dents. (From Vangheluwe, L., *Air-Jet Weft Insertion (Textile Progress)*, The Textile Institute, Manchester, UK, 1999.)

The airflow through the tunnel created by the reed will depend on the following factors:

- Dimension of the reed dents

- Distance between the reed dents

- Angle between the front surface of reed dent and reed channel

The angle between the front surface of reed dent and reed channel prevents divergence of air stream by providing guidance as shown in Figure 8.25.

8.3.5 Relay Nozzles

Relay nozzles are placed across the loom width and they work in conjugation with the profile reed to maintain the velocity of air stream. The purpose of relay nozzle is not to accelerate the pick which is done by the main nozzle. The aim of relay nozzles is to maintain the attained velocity of the pick once the influence of main nozzle diminishes. Relay nozzles move below the shed through the bottom shed line before beat-up. The timing of blowing of relay nozzles is very crucial as it determines the trajectory of the pick as well as air consumption. Around 80% of the compressed air used in air-jet loom is consumed by the relay nozzles. The number of relay nozzles to be used in a loom is dependent on the loom width with wider looms requiring more relay nozzles. Relay nozzles are generally operated in groups so that four or five relay nozzles of a group operate at the same time.

Blowing time of relay nozzles is synchronised with the arrival time of the yarn tip in that particular location as shown in Figure 8.26. The second group of relay nozzles starts blowing of air when the yarn tip arrives at the vicinity. Figure 8.27 depicts that there is certain overlap in the blowing time of various groups of relay nozzles. This overlap should be carefully adjusted so that the length of pick under the influence of relay nozzles does not vary much.

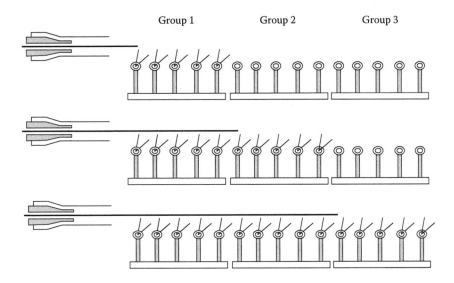

FIGURE 8.26   Timing of relay nozzles.

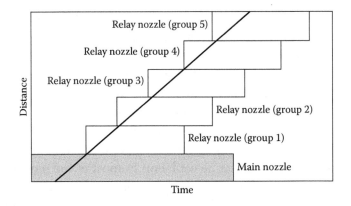

FIGURE 8.27   Synchronisation of relay nozzle timing.

If the number of groups of relay nozzles is more, then their timing can be precisely adjusted to minimise the air consumption without hampering the weft insertion. A simple hypothetical example will make the situation clear. Let it be assumed that there are N relay nozzles which have been divided into n groups. The span of picking is T second. If the number of relay nozzle group is one ($n = 1$), then all the nozzles should be operating for the entire duration of picking. Therefore, the air consumption will be proportional to the product of number of nozzles operating at a time and duration of blowing. In this case, it would be proportional to $N \times T$. If the relay nozzles are divided into two groups, then the number of relay nozzles in each group will be $N/2$. Assuming that the first set of nozzles blow for the entire duration of picking and the second set of nozzles blow only for the second half of the picking span, the air consumption by relay nozzles will be as follows:

$$\frac{N}{2} \times T + \frac{N}{2} \times \frac{T}{2} \text{ (as the second set of nozzles will blow for } T/2 \text{ s)}$$

$$= \frac{3}{4} NT$$

This will yield 25% reduction in air consumption. In the case of three groups of relay nozzles, the total air consumption by relay nozzles will be as follows:

$$\frac{N}{3} \times T + \frac{N}{3} \times \frac{2T}{3} + \frac{N}{3} \times \frac{T}{3} = \frac{2}{3} NT$$

This will yield 33% reduction in air consumption. Therefore, it can be inferred that more number of groups, for a given number of relay nozzles, will be beneficial for reducing air consumption. However, individual control of relay nozzles to minimise the consumption of compressed air seems to be pragmatically difficult due to high capital cost and increased design complexity of the electronic control systems (timers, valves, etc.). Some machine manufacturers (e.g. Tsudakoma) offer individual control of relay nozzles as an optional machine feature. Figure 8.28 shows the grouping of relay nozzles in a Dornier air-jet loom. In most of the modern air-jet looms, the distance between the relay nozzles changes across the loom width. The gap between first few relay nozzles is higher and it reduces at the receiving side of the loom. For example, Sulzer Ruti L5100 air-jet loom

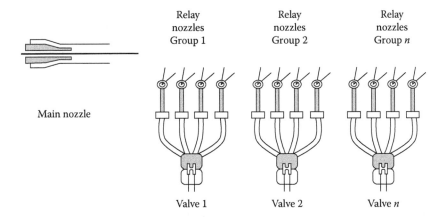

FIGURE 8.28   Grouping of relay nozzles.

with 190 cm width has 28 relay nozzles. The nozzle to nozzle distance for the first 22 relay nozzles is 77 mm, whereas it is only 37 mm for the last six. Besides, in some machines, the first few valves control more number of relay nozzles, whereas the last few values control less number of relay nozzles. All these design aspects ensure precise control over the blowing time of relay nozzles and reduction in the consumption of compressed air.

8.3.5.1 Types of Relay Nozzles

There are two types of relay nozzles: single hole and multihole. In the case of single hole, the cross section can be circular, elliptical or even rectangular. Multihole relay nozzles are also known as shower nozzles. Shower nozzle shows lower variation in the direction of air blowing with change in air pressure. The number of holes and their arrangement in shower nozzle can vary to meet the requirement of weaving. Some of the designs of shower-type relay nozzle are shown in Figure 8.29. It can be seen that a tapered outlet of the relay nozzle maintains the air stream and thus some of the air-jet loom models (Toyota JAT 710) are incorporated with this kind of relay jet design (Figure 8.30).

8.3.6 Design of Main Nozzle

The design of main nozzle is of paramount importance as it influences the initial drag created on weft yarn. Different designs of inserts, through which the yarn is introduced to the tube, are used to meet the requirement. Cylindrical and conical inserts, shown in Figure 8.31, are preferred for the filament and spun yarns, respectively. In the case of cylindrical

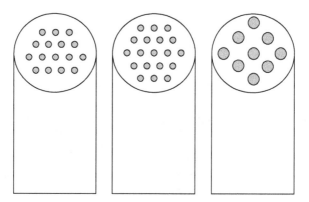

FIGURE 8.29   Different types of multihole or shower nozzle.

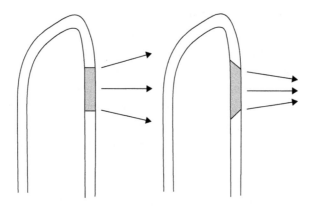

FIGURE 8.30   Horizontal and tapered jets.

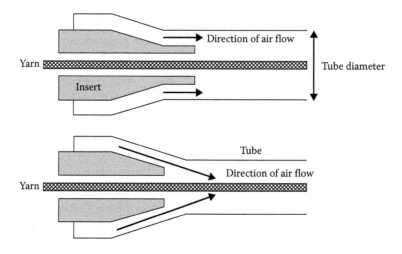

FIGURE 8.31   Main nozzle with cylindrical and conical inserts.

insert, the air flow in the tube is parallel to the yarn axis at the exit point of insert. This creates very gentle action of air on the weft yarn and thus the energy transfer is lower. On the other hand, in the case of conical insert, the air flow in the tube takes place at an angle with the yarn axis at the exit point of insert. This ensures better energy transfer from air to weft yarn. However, the action is harsher in this case as compared to that of parallel insert. The inner diameter of the insert should be higher for coarser yarns. Besides, the diameter and the length of the tube also depend on the yarn linear density. Coarser weft yarns require more drag force for acceleration and thus the length of the tube should be longer in this case.

8.3.7 Weft Storage Systems

Weft storage system is placed between the supply package and main nozzle. The objective of weft storage is to minimise the tension variation in the pick during insertion which would otherwise arise in the absence of this device. There are two types of weft yarn storage systems: loop storage and drum storage. The former is no longer used in modern air-jet looms. Figure 8.32 shows the schematic arrangement of elements in a loop storage system. The yarn is stored in the form of a loop inside the suction tube. The length of the loop depends on the loom width. The system is not convenient for yarns having snarling tendency. Besides, the system does not have adequate control over the yarn during the weft insertion.

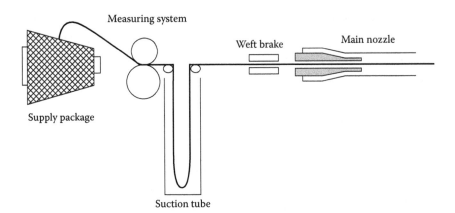

FIGURE 8.32   Loop storage system.

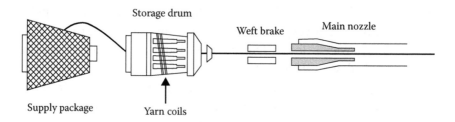

FIGURE 8.33 Drum storage system.

Figure 8.33 shows the drum storage system which accumulates yarns for multiple picks on the drum. The winder of the drum rotates and winds the coils on the drum after unravelling the yarn from the supply package. The drum is made up of multiple polished metallic fingers. The requisite number of coils are unravelled during weft insertion which is generally monitored optically. Then the weft brake acts to stop further withdrawal of coils from drum. The drum storage system exerts good control over the pick during weft insertion as the yarn needs to overcome the friction with the drum during unravelling of yarn coils.

8.3.8 Loom Timing

Figure 8.34 depicts the typical timing of various operations in air-jet loom. The main nozzle is switched on followed by the first set of relay nozzles before the opening of yarn clamp or weft brake. The air flow from the main nozzle straightens the tip of yarn protruding out from the nozzle. The duration of blowing of main nozzle is around 90°–110°. The yarn clamp is closed after the requisite number of coils have been removed from the drum accumulator. The last set of relay nozzles is switched off at last so that the yarn tip is straightened. The following adjustments may be made while setting the timing of relay nozzles of air-jet loom.

- The blowing time of the relay nozzles is delayed for delicate yarns to prevent damage.

- Duration of blowing is increased for continuous filament yarns as compared to equivalent spun yarns as drag force created will be lower in the case of the former.

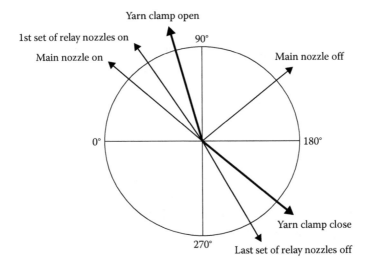

FIGURE 8.34 Timing of operations in air-jet loom.

8.3.9 Effect of Yarn Characteristics on Yarn Velocity

Research work by Adanur and Turel (2004) has shown that the following yarn parameters play a significant role in determining the weft velocity in air-jet weaving.

- Yarn count

- Yarn hairiness

- Yarn bulk

- Yarn twist

- Denier per filament

8.3.9.1 Yarn Count

For a given air pressure, yarn with higher linear density attains lower acceleration and lower velocity in air-jet weaving. Coarser weft yarn experiences more air drag due to higher surface area. However, the mass of a yarn is proportional to its diameter^2. Therefore, the resultant effect of yarn linear density on acceleration can be summarised as follows:

$$\text{Air drag force } \alpha \text{ yarn diameter}$$
$$\text{Yarn mass } \alpha \text{ yarn diameter}^2$$

Therefore

$$\text{Acceleration} = \frac{\text{Force}}{\text{Mass}} \propto \frac{\text{Diameter}}{\text{Diameter}^2} \propto \frac{1}{\text{Diameter}}$$

Therefore, weaving of finer yarn consumes less compressed air in air-jet weaving.

8.3.9.2 Yarn Hairiness

Hairs protruding from the yarn surface increase the friction between air and yarn surface, thus increasing the drag force. Therefore, hairy yarns show higher velocity during air-jet weft insertion. However, higher hairiness may spoil the fabric appearance.

8.3.9.3 Yarn Bulk

For a given combination of air pressure and yarn linear density, bulky yarns will exhibit higher velocity during air-jet weft insertion. This is due to the creation of more drag force on the bulky yarn due to its higher surface area. Therefore, textured yarns show higher velocity than the flat filament yarns during air-jet weft insertion.

8.3.9.4 Yarn Twist

Increase in twist in spun yarn causes reduction in yarn diameter and hairiness. Lower diameter and reduced hairiness both decrease the drag force acting on the yarn. Therefore, the yarn velocity reduces with the increasing yarn twist.

8.3.9.5 Denier per Filament

For a given total yarn denier, lower denier per filament leads to higher weft velocity. The total surface area of yarn increases as the denier per filament reduces for a constant yarn denier. Thus the drag force acting on the yarn increases, resulting in higher weft velocity.

8.3.10 Air Index

The air index (AI) indicates the air-friendliness of a yarn and its weavability on air-jet looms (Wahhoud et al., 2004). Air index is measured by Air Index Tester developed by Picanol (Belgium). The schematic arrangement of the air index tester is shown in Figure 8.35. The weft

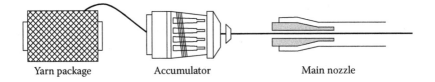

Yarn package   Accumulator   Main nozzle

FIGURE 8.35   Schematic representation of air index tester.

accumulator winds defined number of coils ensuring that all the coils have the same diameter. The calibrated main nozzle is equipped with air pressure regulating system.

The equipment determines the speed of a pick accelerated by the main nozzle at a specific air pressure. The air index (m/s) provides an indication about the maximum weft insertion speed that can be attained by a particular yarn and the required air pressure for this at the main nozzle. Thus it implies the productivity as well as energy requirement during the weaving of a particular yarn. A high air index value implies higher achievable weft insertion speed and lower consumption of compressed air. The air index coefficient of variation (CV%) indicates the speed variation in weft with the length of yarn tested. The implication of air index CV% is as follows:

- >5%: Poor weavability

- 3%–5%: Reasonable weavability

- <3%: Weavable without significant problems

Research work has shown that air index increases with the increase in yarn hairiness as the drag experienced by the yarn increases. On the other hand, air index reduces with the increase in yarn twist as the surface area of yarn reduces due to increased compactness of fibres inside the yarn structure. In an experiment, air index showed good association with compressed air consumption during weaving in actual conditions (190 cm loom width and 900 picks per minute loom speed). The air-jet weaving machine was equipped with a main nozzle regulating system to ensure constant arrival time of pick at the receiving end of the loom. It was also found that higher air index causes reduction of air pressure at main nozzle and vice versa.

8.3.11 Tension Profile of Weft Yarn

Figure 8.36 shows a typical tension profile of weft yarn in air-jet weaving. As the blowing of compressed air starts from the main nozzle, tension generates on the weft yarn. There is random fluctuations in tension due to unravelling of yarn coils from the drum accumulator and formation of balloon between the drum accumulator and the main nozzle. However, the main tension peak generates as soon as the weft brake is applied after the removal of requisite number of yarn coils. At this point, the entire kinetic energy of the moving yarn is converted to strain energy. The peak tension generated in the weft yarn becomes higher if the yarn is coarser and its velocity is more.

If T, L and v are the linear density (in direct system), length and velocity of the pick, respectively, then

$$\text{Kinetic energy of the yarn} = \frac{1}{2}TLv^2 \tag{8.8}$$

$$\text{Strain energy} = \int_0^\varepsilon F \, d\varepsilon$$

where F is the force acting on the yarn at elongation ε.

For linear stress–strain relationship, $F = YT\varepsilon/L$.

where Y is the Young's modulus of yarn.

$$\text{So strain energy} = \frac{YT}{L} \int_0^\varepsilon \varepsilon \, d\varepsilon = \frac{YT\varepsilon^2}{2L}. \tag{8.9}$$

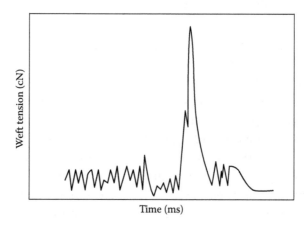

FIGURE 8.36 Tension profile of weft yarn.

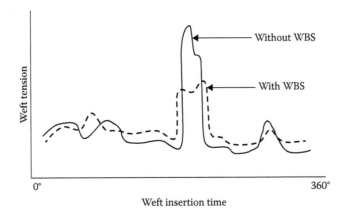

FIGURE 8.37   Reduction of peak tension by weft braking system.

Equating kinetic and strain energy, the following expression is obtained.

$$\varepsilon = \frac{Lv}{\sqrt{Y}} \qquad (8.10)$$

Now

$$\text{Force} = \frac{YT\varepsilon}{L} = \frac{YT}{L} \times \frac{Lv}{\sqrt{Y}} = \sqrt{Y}Tv \qquad (8.11)$$

This expression clearly shows that the peak tension generated in the weft yarn becomes higher if the yarn is coarser and its velocity is more.

The problem of tension peaks can be mitigated by using capstan-type weft braking system (WBS) or automatic braking system (ABS). This brake creates additional frictional resistance on the weft yarn towards the end of weft insertion. Figure 8.37 depicts the tension profile of weft yarn with WBS or ABS used in Tsudakoma ZAX 9200 and Toyota JAT 710 air-jet looms, respectively. The tension peak reduces significantly with this system. The system is more beneficial for the wide width loom as the peak tension is higher.

8.3.12 New Features in Air-Jet Looms

8.3.12.1 Programmable Speed Control

In most of the modern air-jet looms, the loom speed (picks per minute) can be programmed at different levels for different sections of the fabric. When different types of weft yarns are used for weaving, the loom speed

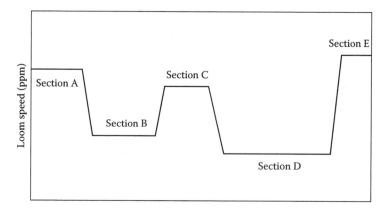

FIGURE 8.38   Programmable speed control.

is limited by the weaving performance of the weakest weft yarn. However, with programmable speed control feature, different speeds can be set for different sections of the fabric (Figure 8.38). Therefore, when stronger weft is used, loom speed is increased and vice versa. So the overall productivity of the loom increases.

8.3.12.2 Individual Control of Relay Nozzles

Generally a group of relay nozzles are controlled by a single valve. So all the relay nozzles of a group are activated at the same time, precluding the possibility of precise control of timing of individual nozzles. This can cause significant wastage of compressed air. Some new models of air-jet looms have provision for individual control of relay nozzles with dedicated valves. Figure 8.39 depicts one such system offered by Tsudakoma (model ZAX 9200).

8.3.12.3 Adaptive Control System

Adaptive control system in air-jet loom helps to reduce the air consumption (Shyam, 2014). The waviness of yarn, especially for filament yarns, increases from surface to core of the weft package. Therefore, the drag force acting on the pick also increases as the weft package diameter reduces. Thus the time required by the pick to travel through the shed also reduces progressively. When actual arrival time (T_a) of the pick becomes earlier than expected arrival time (T_e), air consumption can be reduced by the following ways.

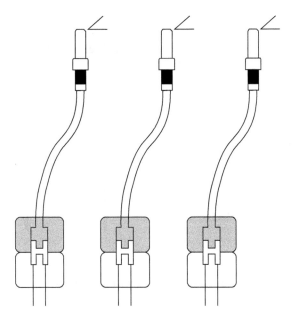

FIGURE 8.39 Individual control of relay nozzles.

- Delaying the valve opening time of main nozzle
- Increasing the loom speed
- Reducing the air flow

Figure 8.40 depicts the change in arrival time of the pick at the receiving end of loom without any control system. As the diameter of weft package reduces, the pick arrives earlier than expected, that is $T_a < T_e$. If the timing of opening of main nozzle valve is delayed by ΔT, then the pick will reach the receiving end of loom exactly at the expected time. This is indicated pictorially in Figure 8.41.

In Figure 8.41, the distance–time relationship of the pick before and after the adjustment of timing of main nozzle valve is shown with broken and solid lines, respectively. When the timing of opening of main nozzle valve is delayed by ΔT, *that is* $(T_e - T_a)$, the actual arrival time of pick is also delayed by ΔT. So the actual arrival time coincides with the expected arrival time.

Figure 8.42 shows the principle of reduction of air consumption by increasing the loom speed. As the loom speed increases, the time available to complete one pick cycle (360°) reduces. Let us assume that the loom is

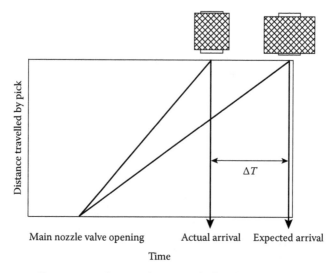

FIGURE 8.40 Change in pick arrival time with the change in package diameter.

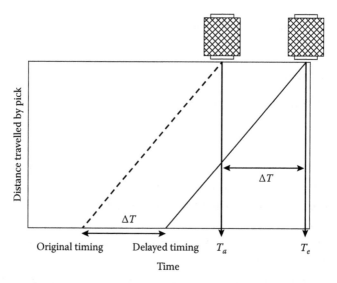

FIGURE 8.41 Delaying the timing of opening of main nozzle valve.

running at 500 rpm. The expected and the actual durations of pick insertion are equivalent to 120° and 100° rotation of loom shaft, respectively. So the actual arrival time of the pick at the receiving end of loom precedes the expected arrival time by 20°. If the loom speed is now increased to 600 rpm, then the mismatch between the expected and actual arrival times can be eliminated. This would happen as 100° and 120° rotations of

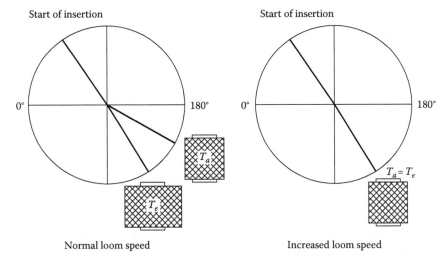

FIGURE 8.42 Increasing weaving speed for adaptive control.

loom shaft with loom speed of 500 and 600 rpm, respectively, lead to the same time as shown in the following text.

$$\text{Time for peak insertion (s)} = \frac{60}{\text{Loom rpm}} \times \frac{\text{Span of weft insertion (in deg)}}{360°}$$

$$= \frac{60}{500} \times \frac{100°}{360°} = 0.033 \text{ s (for 500 rpm)}$$

$$= \frac{60}{600} \times \frac{120°}{360°} = 0.033 \text{ s (for 600 rpm)}$$

8.4 WATER-JET PICKING SYSTEM

In water-jet loom, the pick is inserted by a tiny jet of water emanating at a very high speed from a nozzle. The densities of air and water are 0.001293 and 1 g/cm³, respectively. Thus the density of water is around 800 times higher than that of air. Therefore, from the expression of drag force (Equation 8.5), it can be inferred that the drag created on the pick by the water-jet is much higher than that of air-jet. Though the fundamental principle of weft insertion is similar in air-jet and water-jet looms, there are some differences as well. Only one nozzle, that is main nozzle, is used in water-jet loom and there is no relay nozzle. The system is suitable for hydrophobic yarns as hydrophilic yarns may pick up water and this may alter the nature of weft insertion. It has to be ensured that sizing materials used in warp yarns remain insoluble in water. The quality requirement of water to be used for water-jet weaving is very stringent, which will be discussed in the next section. Due to these limitations, the popularity of water-jet loom is quite limited. Power consumption in water-jet loom per unit area of fabric is the lowest among the commercially successful systems.

8.4.1 Water Quality

The water to be used in water-jet loom has very stringent quality requirement. Inferior water quality can cause scale, rust, corrosion, blockage of filter and stains in fabric. Some of the quality requirements are given in Table 8.3 (Technical literature of Toyota water-jet loom LWT 710, 2015).

The flowchart of water treatment for water-jet weaving is shown in Figure 8.43. Activated charcoal is used to remove organic matter from the water. Softening of water is done to remove the calcium and magnesium salts. Fine filtration is done to remove solid particles.

TABLE 8.3   Water Quality for Water-Jet Looms

Water Quality	Allowable Standard	Problem Caused
Turbidity	<2 ppm	Scale, rust, corrosion, filter blockage, etc.
Total hardness (calcium and magnesium salts)	<3 ppm	Scale in nozzle
Total iron (Fe) and manganese (Mn)	<0.2 ppm	Corrosion and colouration
Chloride ion	<20 ppm	Corrosion
pH	6.5–7.5	
Temperature	14°C–20°C	High temperature may dissolve the size

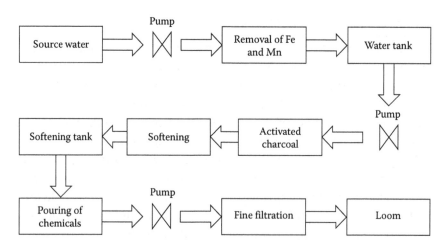

FIGURE 8.43   Flowchart of water treatment.

TABLE 8.4   Water Consumption in L/h at Different Loom Speeds

Water Requirement per Pick (cm³)	Loom Speed (Picks per Minute)		
	600	800	1000
2.4	86.4	115.2	144
3.0	108	144	180
3.75	135	180	225

8.4.2 Water Consumption

The water consumption in a water-jet loom can be estimated using the following expression.

$$V(\text{L/h}) = \frac{\pi \times D^2 \times L}{4} \times N \times 60 \times 10^{-6} \qquad (8.12)$$

where

D is plunger diameter (mm)
L is plunger stroke (mm)
N is loom speed (picks per minute)

The water requirement per pick can vary from 2.4 to 6.4 cm³. The water consumption in litre per hour for different loom speed is shown in Table 8.4.

8.4.3 Loom Timing

Figure 8.44 depicts a typical timing of primary motions in a water-jet loom. The cam-driven sley has a substantial dwell period. The weft clamp

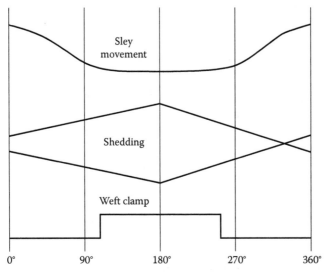

FIGURE 8.44   Timing of primary motions of water-jet loom.

opens during this dwell period and thus the pick is inserted. The shed becomes completely open at 180° position of the loom shaft and it closes before the beat-up. There is no dwell period for the shed which can be understood from continuous change of positions of top and bottom shed lines. Weaving without any dwell period of shed is possible as very low shed depth is enough for the passage of the pick.

8.5 RAPIER PICKING SYSTEM

The word 'rapier' means a thin and sharp pointed sword. In rapier weaving, pick is inserted with the help of one or two rapiers. Rigid rapier system was invented in 1870 and was further developed by O. Hal-lensleben in 1899. The principle of loop transfer was developed by John Gabler in 1922. The idea of tip transfer at the middle of shed was developed by R. Dewas in 1939.

Rapier is a very versatile weft insertion system (Figure 8.45). While projectile and air-jet picking systems are suitable for simple fabrics in high and low loom width, respectively, rapier system can weave complex fabrics. The weft yarns having linear density ranging between 1 and 4500 tex can be handled using a rapier system, which is capable of handling high-performance yarns like para-aramid, carbon and glass and also very delicate yarns like silk. Rapier weaving is popularly used to weave industrial fabrics which often require very coarse and heavy pick to be inserted through the shed.

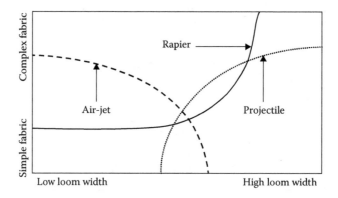

FIGURE 8.45 Application domains of shuttleless looms.

8.5.1 Classification of Rapier Picking System

Rapier picking system can be classified as shown in Figure 8.46. In the case of single rapier machine, the rapier travels from one side of the loom to the other and picks up the weft yarn. Then the rapier completes its return movement and inserts the pick through the shed as shown in Figure 8.47. As a single rapier covers the entire width of the loom, it must have sufficient rigidity to avoid any bending and buckling. Thus, single rapiers are invariably rigid, which means that the rapier rod is made of either metal or composite. Rigid rapiers do not need any guidance during weft insertion. There is no need to transfer the pick at the middle of the shed. Therefore, a single rapier system is advantageous for weaving weft yarns which are difficult to grip or exchange.

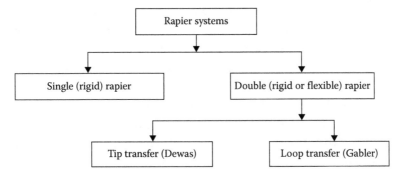

FIGURE 8.46 Classification of rapier picking systems.

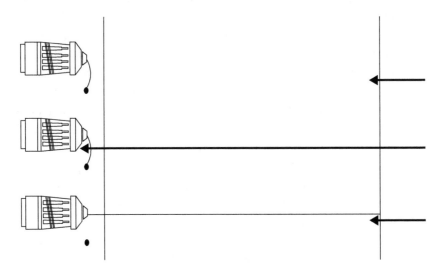

FIGURE 8.47 Steps of weft insertion by rigid rapier.

In single rapier, half of the rapier movement is wasted as it does not contribute to weft insertion. This may limit the loom speed as the rapier has to travel twice the loom width for the insertion of one pick. For example, if the loom width is 2 m and average velocity of the rapier is 40 m/s, then the rapier will take 0.1 s to complete its to-and-fro movement. If a double rapier system is used, then each one of them travels up to the middle of shed and then returns. Therefore, the total distance travelled by each rapier is 2 m. Considering the same average velocity of rapier, the time requirement for rapier movement becomes 0.05 s. Therefore, a double rapier system presents possibilities to attain higher loom speed. Besides, being rigid, single rapier systems require more space and energy. Therefore, their popularity is quite limited.

The problem of single rapier can be mitigated by using two rapiers. In a double rapier system, one rapier brings the pick-up to the middle of the shed and then the other rapier completes the remaining part of weft insertion. As each of the two rapiers extends only up to the half of the loom width, double rapiers can be either rigid or flexible. When flexible rapiers are used, retracting rapiers can be wound on a drum reducing the space requirement. Therefore, double flexible rapiers are more common than the double rigid rapiers. Flexible rapiers need guiding device (ribbed band guide) when they travel through the shed. This guidance is provided by the ribbed guide which remains within the shed during weft insertion and comes out of the shed during beat-up. During the movement of the rapier, the reed and the ribbed guide assembly remain stationary for unobstructed movement of the former. Use of delicate warp yarns becomes a bit difficult in this case due to continuous abrasion between the warp and the guide.

Based on the mode of pick transfer at the middle of the shed, double rapier systems can be of two types: tip transfer (Dewas) and loop transfer (Gabler). The heads of giver and taker rapiers for tip transfer system or Dewas system are shown in Figure 8.48. Both the rapier heads have yarn clamps for holding the pick securely. Under normal circumstances the clamps remain closed. Each clamp has one clamp opener. When the clamp opener is pressed, the clamp opens. For example, the clamp opener of taker rapier is pressed by a cam fixed on the sley before the tip transfer. This facilitates the gripping of yarn tip by clamp of taker rapier. The clamp opener of taker rapier is pressed again by another cam outside the shed when the pick insertion is over.

In the case of tip transfer system or Dewas system, the giver rapier grips the tip of the pick and carries it up to the middle of the shed. Then the tip is transferred from the giver rapier head to the taker rapier head (Figure 8.49). The latter then pulls the pick to the other side of the loom.

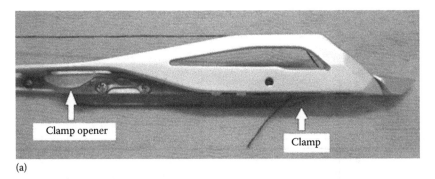

(a)

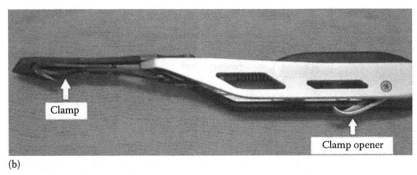

(b)

FIGURE 8.48 (a) Giver rapier head and (b) taker rapier head.

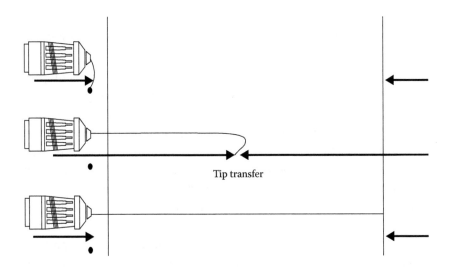

Tip transfer

FIGURE 8.49 Steps of weft insertion by double rapier (Dewas system).

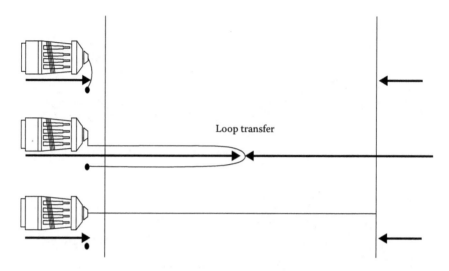

FIGURE 8.50   Steps of weft insertion by double rapier (Gabler system).

In contrast, in the case of loop transfer system or Gabler system, the giver rapier carries the weft yarn in the form of a loop up to the middle of the shed (Figure 8.50). This loop is formed as the tip of the weft yarn is gripped at a point, at the side of weft accumulator, when the giver rapier comes in contact with the weft. The tip of the weft remains gripped at the gripping point till the giver rapier travels up to the middle of shed. After this, the taker rapier just pulls the loop straight to complete the weft insertion.

8.5.2 Displacement, Velocity and Acceleration Profiles of Rapier
8.5.2.1 Displacement
Figure 8.51 depicts a typical displacement profile of giver and taker rapiers. Here it has been assumed that both the rapiers are following simple harmonic motion in their displacement. The giver rapier starts its movement from one side (generally left-hand side) of the loom, while the taker rapier does it from the opposite side. So, at the beginning, the distance between the giver and taker rapier heads is approximately equal to loom width. The distance between the two heads reduces as both of them move towards the middle of the shed. When they meet at the middle of the shed, either loop or tip transfer takes place. The displacements of the giver and taker rapiers overlap a bit to facilitate this transfer. Therefore, the displacement profile of the pick follows those of giver rapier (solid line) and taker rapier (broken line) in the first half and second half, respectively, of its movement.

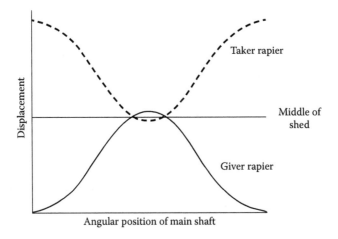

FIGURE 8.51   Displacement profiles of rapiers.

8.5.2.2 Velocity

The velocity profiles of giver and taker rapiers are shown in Figure 8.52 by solid and broken lines, respectively. The two rapiers always move in opposite directions which can easily be understood from their velocity profiles. Initially the velocity of both the rapiers increases with time. The peak velocities are attained when one-fourth of the total time for movement elapses. After this, the velocity of both the rapiers starts to decrease. The velocity of rapiers becomes zero when half of the total time for

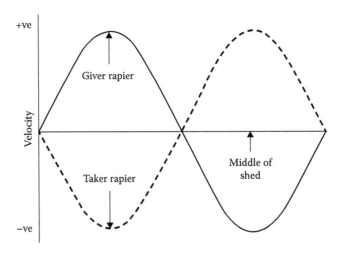

FIGURE 8.52   Velocity profiles of rapiers.

movement elapses, that is the rapier heads reach the middle of the shed. The control of the pick gets transferred from the giver rapier to the taker rapier at this point. After this, both the rapiers accelerate again and attain the peak velocities, albeit in opposite directions as compared to the previously stated peak velocities, when three-fourth of the total time for movement elapses. Then both the rapiers decelerate and come to rest at the same position from where they started their respective movements.

8.5.2.3 Acceleration

Figure 8.53 shows the acceleration profiles of rapiers. The two rapiers follow simple harmonic motion. Therefore, they always have finite acceleration towards their respective middle point of path or sweep. The magnitude of acceleration is proportional to the displacement of the rapier head from the middle point of their path. This can be better understood from the schematic representation of Figure 8.54. For the giver rapier, the left-hand side edge of the loom and middle point of the shed are the two extreme points of motion. On the other hand, for the taker rapier, right-hand side edge of the loom and middle point of the shed are the two extreme points of motion. The magnitude of acceleration is the maximum at the extreme points of motion, that is at the sides and at the middle of shed. It is important to note that when the pick transfer takes place at the middle of shed, not only the acceleration of the rapier heads are at maximum level but they are also at the opposite directions. This may cause significant rise in tension level of the pick.

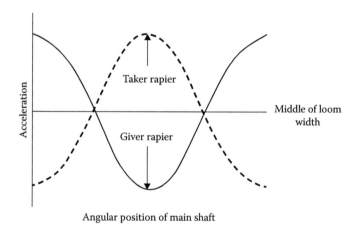

FIGURE 8.53 Acceleration profiles of rapiers.

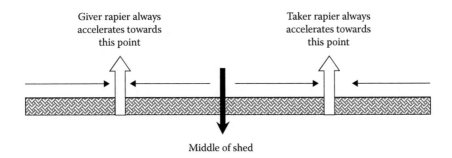

Giver rapier always
accelerates towards
this point

Taker rapier always
accelerates towards
this point

Middle of shed

FIGURE 8.54   Simple harmonic motions of two rapiers.

8.5.3 Tip Transfer or Dewas System

Figure 8.55 depicts the stages of tip transfer in Dewas system. The open clamp of the giver rapier grips the yarn presented by the weft selecting finger before entering the shed. After controlled closure of this clamp, scissors cut the selected weft yarn at the fabric side. The giver

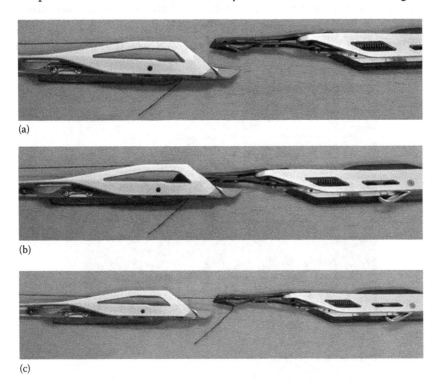

(a)

(b)

(c)

FIGURE 8.55   (a) Rapier heads before tip transfer, (b) during tip transfer and (c) after tip transfer.

rapier moves from left to right with the tip of pick gripped firmly. The taker rapier reaches the middle of the shed a bit early. Then a fixed cam mounted on the sley opens the clamp of the taker rapier. The heads of the two rapiers slide past each other and in the process the tip of pick comes under the control of the taker rapier clamp. The clamp of the taker rapier closes to secure the tip of pick and then the clamp of the giver rapier opens due to the action of another cam fixed on the sley. Then both the rapiers retract and withdraw from the shed. When the taker rapier comes out of the shed, its clamp opens again by the action of a cam and the pick gets released. The pick is then beaten up by the reed while the shed closes.

8.5.4 Multicolour Weft Selection

The velocity of the giver rapier is relatively slower at the start of its motion. Thus the rapier gets sufficient time to grip the yarn. Moreover, the weft yarn of requisite colour or type (fibre, linear density, etc.) is positioned in the path of the rapier by the weft finger. Therefore, use of multi-colour weft is relatively easier in rapier loom. In some machines, up to 16 colours can be used in weft. Figure 8.56 shows the weft selection mechanism for

FIGURE 8.56   Weft selecting system.

multiple colours. Depending on the pick sequence, a particular electro-magnet is activated which then pulls the corresponding tongue through a carriage. The corresponding weft finger is lowered in the position suitable for presenting the weft to the rapier head.

8.6 SELVEDGE FORMATION IN SHUTTLELESS LOOMS

In shuttleless looms picks are inserted from one side of the loom. After the insertion of one pick, it is cut near the two selvedges. Thus small length of weft yarn remains freely hanging at the two sides of the fabric. If the selvedges are not formed then the ends may come out of the fabrics during subsequent processing. Therefore, shuttleless looms are provided with selvedge forming mechanisms.

Tucked selvedge can be used in projectile and air-jet looms, whereas leno selvedge can be used in air-jet or rapier looms. In projectile looms, tucked selvedge is formed with the aid of a tucking needle as shown in Figure 8.57. The tucking needle enters the shed from under the bottom shed line and pulls in the free end of the pick inside the next shed. In air-jet looms, additional air blowers are installed near the selvedge. These blowers blow the free end of the pick into the shed.

Leno selvedges are made by using two small packages of warp yarns mounted on a disc which rotate either by 180° or by 360° in every pick (Figure 8.58). The former produces half-cross leno and the latter produces full-cross leno. Half-cross leno is suitable for spun yarns, whereas full-cross leno is preferred for smoother yarns as the latter offers tighter grip on the picks. Fused selvedge is produced for thermoplastics yarns.

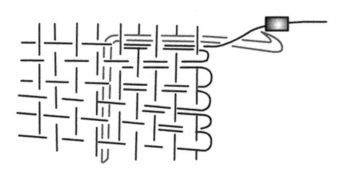

FIGURE 8.57 Formation of tucked selvedge. (From Banerjee, P.K., *Principles of Fabric Formation*, CRC Press, Boca Raton, FL, 2015.)

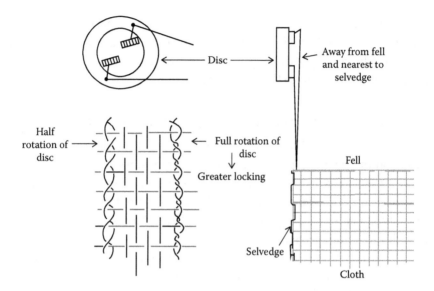

FIGURE 8.58 Formation of leno selvedge. (From Banerjee, P.K., *Principles of Fabric Formation*, CRC Press, Boca Raton, FL, 2015.)

8.7 MULTIPHASE WEAVING

Multiphase loom is being discussed within the category of shuttleless looms as the weft insertion in multiphase loom is done without using a shuttle. However, circular looms and ripple shed looms which employ shuttle also comes under multi-phase weaving. One of the ways to express the production capacity of a loom is weft insertion rate. The weft insertion rate is expressed as follows:

$$\text{Weft insertion rate (m/min)} = \text{Picks per minute} \times \text{reed width (m)}$$

As the loom width increases, the weft insertion takes more time and thus the loom speed (picks per minute) reduces as shown by solid line in Figure 8.59. However, the weft insertion rate (shown by broken line) increases with the increase in loom width. In single-phase weaving machines, the picks are accelerated and decelerated within a very small span of time. Therefore, the stresses acting on the yarn and various components of the machine become very high. This is one of the major limitations of single-phase weaving machines. The maximum weft insertion rate, as stated in the technical literature of loom

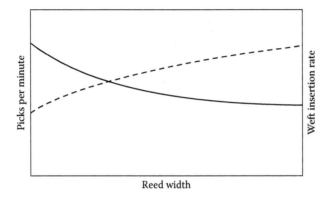

FIGURE 8.59   Influence of loom width on picks per minute and weft insertion rate.

manufacturers, may not be true for all types of yarn. Table 8.5 presents some relevant data in this regard.

From Table 8.5, it is seen that the weft insertion rate can be increased drastically in multiphase weaving machines. It was first shown by Sulzer Textil (model M8300) in ITMA 1995. Figure 8.60 shows the schematic view of shed formation in multiphase loom.

TABLE 8.5   Technical Specifications of Some Weaving Machines

Weaving Machine	Weft Insertion Rate (m/min)	Maximum Width (cm)	Yarn Count Range (tex)
Sultex projectile P7300	1500	390	
Dornier air-jet	2650	540	6–100 cotton 5–220 polyester (smooth) 5–110 polyester (textured)
Picanol air-jet OMNIplus		400	6–250 (spun) 1.1–110 (filament)
Tsudakoma water-jet ZX8100		230	
Toyota water-jet LWT710	2500	230	
Dornier rapier P_1	1200	430	0.8 (silk)–4500 (glass)
Picanol rapier Opti Max-i	1700	540	5–330 (spun) 2.2–330 (filament)
Sulzer Textil multi-phase M8300	6088	188.5	15–60

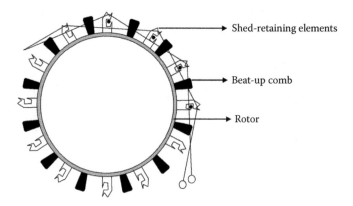

FIGURE 8.60   Principle of shed formation in multiphase weaving.

Multiple sheds are created simultaneously by using shed-retaining elements which are mounted on a rotating rotor. The shed-retaining elements, which have profiled shape, are arranged in several rows (12) across the width of the rotor. Shed-retaining elements control the ends for forming the sheds. A slot is present within each shed-retaining element to create a channel for weft insertion. The beat-up is done by the beat-up combs which are also arranged in several rows across the width of the rotor. The rows of shed-retaining elements and beat-up combs are arranged alternately. Four sheds are created simultaneously and thus four picks are inserted at a time, in staggered fashion, from four weft packages. The picks are inserted by compressed air and assisted by relay nozzles which are integrated in the shed-forming elements at some intervals. Once the pick insertion is over, the picks come out through the slots of shed-retaining elements. Then beat-up combs push the newly inserted picks to proper position. The weft insertion rate can go up to 6088 m/min as a multiphase loom of 188.5 cm width can run at a speed of 3230 picks/min. Multiphase weaving system creates very low tension in the pick and is suitable for weaving standard fabrics having specified yarn counts and fabric sett. Figure 8.61 presents a comparison between a single-phase air-jet weaving machine and a multiphase weaving machine in terms of weft insertion rate and velocity of individual pick. The broken and solid lines indicate the pick velocity and weft insertion rate (m/min), respectively. It is evident that the multiphase weaving provides higher weft insertion rate even with lower pick velocity.

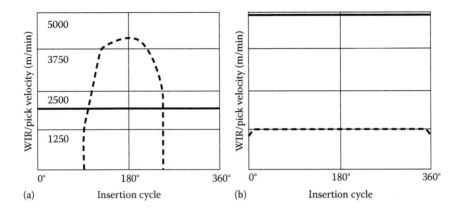

FIGURE 8.61   Weft insertion rate and pick velocity in (a) single-phase (air-jet) and (b) multiphase weaving.

The limitations of multiphase weaving machine described here are as follows:

- Only plain and 2/1 twill weaves can be woven.
- End density cannot be changed as it is dependent on the number of shed-retaining elements in a row.
- Pick density is low due to low beat-up force.

Because of the aforesaid reasons, the technology has not seen much success in the industry. Sulzer Textil has not demonstrated any development in multi-phase weaving technology in ITMA 2007 and 2011.

NUMERICAL PROBLEMS

8.1   Two cylinders of identical material have the same length and mass. One cylinder is solid and the other one is hollow having outer radius twice the inner radius. Compare their torsion rigidities.

Solution:
Let the length of the cylinders be l.

Radius of solid cylinder is r.

Outer and inner radii of hollow cylinder are $2r_1$ and r_1, respectively.

As the mass, length and density of material are same for the two cylinders, the area of their cross section must be same.

So

$$\pi r^2 = \pi\left[(2r_1)^2 - r_1^2\right] = \pi 3r_1^2$$

or

$$r^2 = 3r_1^2$$

Now

Torsional rigidity of solid rod

$$= \tau = \frac{\pi G r^4}{2l} \text{ (as } \theta = 1 \text{ radian for torsional rigidity)}.$$

So

$$\frac{\tau_{solid}}{\tau_{hollow}} = \frac{\dfrac{\pi G r^4}{2l}}{\dfrac{\pi G\left[(2r_1)^4 - (r_1)^4\right]}{2l}}$$

$$= \frac{r^4}{15r_1^4} = \frac{\left(3r_1^2\right)^2}{15r_1^4} = \frac{9r_1^4}{15r_1^4} = \frac{3}{5}$$

So the ratio of torsional rigidities of solid and hollow cylinders is 3:5.

8.2 A couple of 65×10^{10} dyne-cm was applied at the free end of the torsion rod having 80 cm length and 2 cm diameter. The rod was twisted through 90° angle. Calculate the modulus of rigidity of the material.

Solution:
Length (l) = 80 cm, radius (r) = 2/2 = 1 cm.
   Twist of the rod (θ) = 90° = $\pi/2$ rad
   Couple (T) = 65×10^{10} dyne-cm.
   Now

$$T = \frac{\pi \theta G r^4}{2l}$$

or

$$65 \times 10^{10} = \frac{\pi \times \dfrac{\pi}{2} \times G \times 1^4}{2 \times 80}$$

or

$$G = 2.11 \times 10^{13} \text{ dyne/cm}^2.$$

So, modulus of rigidity of the material is 2.11×10^{13} dyne/cm^2.

8.3 The area moment of inertia of cross section (about diameter) of a torsion rod is 11.05 cm^4. The length of the rod is 0.8 m. The modulus of rigidity of the material is 8000×10^7 N/m^2. Calculate the power required for picking if the loom speed is 300 picks per minute and twist angle of the torsion rod before picking is 30°.

Solution:

Area moment of inertia (MI) of circular cross section about diameter (used in bending deformation)

$$= \frac{\pi d^4}{64} = \frac{\pi r^4}{4}$$

Polar MI (J) of circular cross section about the axis passing through center (used for torsion)

$$= \frac{\pi d^4}{32} = \frac{\pi r^4}{2}$$

Here

$$\frac{\pi d^4}{64} = 11.05 \text{ cm}^4 \left(\text{area MI about diameter} \right).$$

So

$$\text{Polar MI}(J) = 2 \times \frac{\pi d^4}{64} = 22.1 \text{ cm}^4 = 22.1 \times 10^{-8} \text{ m}.$$

Length of the rod (l) = 0.8 m

Here

$$\text{Twist angle of the rod} = \theta = 30° = \frac{\pi}{6}\,\text{rad.}$$

It is known that

$$\frac{T}{J} = \frac{G\theta}{l} \tag{8.13}$$

Work done can be calculated from the area under the torque (T)–twist angle curve. For linear relationship

$$\text{Work done/pick} = \frac{T\times\theta}{2} = \frac{J\times G\times\theta}{l}\times\frac{\theta}{2} = \frac{JG\theta^2}{2l}$$

$$\text{Work done/pick} = \frac{22.1\times10^{-8}\times8000\times10^7}{2\times0.8}\times\left(\frac{\pi}{6}\right)^2 = 3029.4\,\text{J}$$

$$\text{Number of picks inserted in 1 s} = \frac{300}{60} = 5.$$

$$\text{Work done/s} = 3029.4 \times 5\,\text{J/s} = 15147\,\text{W} = 15.1\,\text{kW.}$$

So

$$\text{Power required for picking} = 15.1\,\text{kW.}$$

8.4 The length and diameter of a torsion rod is 72 and 1.5 cm, respectively. The modulus of rigidity of the material is 8.3×10^{10} N/m². If the mass of the projectile is 40 g and its velocity while leaving the picker is 24 m/s, then calculate the fraction of energy utilised during picking considering that the twist angle of the torsion rod before picking is 30°.

Solution:
Modulus of rigidity $(G) = 8.3 \times 10^{10}$ N/m².
 Diameter of rod $(d) = 1.5$ cm $= 0.015$ m and length $(l) = 72$ cm $= 0.72$ m.
 Twist angle of the rod $\theta = 30° = \pi/6$ rad $= 0.523$ rad.

So

$$\text{Polar moment of inertia} = J = \frac{\pi d^4}{32}$$

$$= \frac{\pi \times 0.015^4}{32} = 4.97 \times 10^{-9} \text{ m}^4$$

$$\text{Torque } T = \frac{J \times G \times \theta}{l} = \frac{4.97 \times 10^{-9} \times 8.3 \times 10^{10} \times 0.523}{0.72}$$

$$= 299.6 \text{ N} \cdot \text{m}.$$

$$\text{Work done in twisting the torsion rod} = \frac{T \times \theta}{2} = \frac{299.6 \times 0.523}{2} = 78.34 \text{ J}.$$

$$\text{Kinetic energy of the projectile} = \frac{1}{2} mv^2 = \frac{1 \times 0.04 \times 24^2}{2} = 11.52 \text{ J}.$$

Fraction of energy utilisation = 11.52/78.34 = 14.7%.

So fraction of energy utilised during picking is 14.7%.

8.5 A polyester weft yarn of 2 m length and 20 tex linear density was moving at a velocity of 20 m/s when the weft brake attached with the weft accumulator of air-jet loom was activated. If all the kinetic energy was used for extending the weft, then calculate the resulting elongation. Also calculate the peak force acting on the yarn as a result of energy transformation. Young's modulus of the yarn is 200 g/denier.

$$\text{Modulus in GPa} = \frac{\text{Modulus in g/denier} \times \text{density of fibre (g/cm}^3)}{11.33}$$

Solution:
Length of pick is 2 m.
So

$$\text{The mass of pick} = \frac{20}{1000} \times 2 = 0.04 \text{ g} = 4 \times 10^{-5} \text{ kg}$$

Let the area of cross section of fibres in yarn = A cm². Density of polyester fibre is 1.38 g/cm³.

So

$$A \times 1000 \times 100 \times 1.38 = 20$$

or

$$A = \frac{20}{1.38 \times 1000 \times 100} = 1.45 \times 10^{-4} \text{ cm}^2 = 1.45 \times 10^{-8} \text{ m}^2$$

Now

Young's modulus of yarn in GPa

$$= \frac{\text{g/denier} \times \rho(\text{g/cm}^3)}{11.33} = \frac{200 \times 1.38}{11.33} \text{ GPa}$$

$$= 24.36 \times 10^9 \text{ Pa}.$$

If the increase in length of the pick is l m, then equating kinetic energy with the strain energy:

$$\frac{1}{2} m v^2 = \frac{1}{2} \times \frac{YAl^2}{L}$$

or

$$\frac{1}{2} \times 4 \times 10^{-5} \times 20^2 = \frac{1}{2} \times 24.36 \times 10^9 \times 1.45 \times 10^{-8} \times \frac{l^2}{2}$$

or

$$l = 9.5 \times 10^{-3} \text{ m} = 9.5 \text{ mm}.$$

Now

$$\text{Force } (F) = \frac{YAl}{L} = 24.36 \times 10^9 \times 1.45 \times 10^{-8} \times 9.5 \times 10^{-3} \times \frac{1}{2} = 1.68 \text{ N}.$$

So, elongation and peak force are 9.5 mm and 1.68 N, respectively.

8.6 Water consumed per pick in a water-jet loom running at 600 picks per minute is 2 mL. The water-jet acts for 26° rotation of loom shaft. The outer and inner diameters of the jet at the entry side are 8 mm and 4 mm, respectively. If the outer and inner diameters of the jet at the exit (atmospheric pressure) are 4 and 2 mm, respectively, then calculate the water-jet velocity at the exit assuming streamline flow. What should be the pressure at the supply side of the water-jet?

Solution:

Here

$$\text{Time required for one pick cycle is} = \frac{60}{600} = 0.1 \text{ s}$$

The span of jetting is 26° of rotation of loom shaft.

So

$$\text{Time of jetting} = \frac{26}{360} \times 0.1 \text{ s} = 7.2 \times 10^{-3} \text{ s}.$$

Now, in 7.2×10^{-3} s water output is 2 mL.

In 1 s water output will be 277 mL or 277×10^{-6} m³.

Let the velocity of the water-jet at the exit be V_2 m/s.

So

$$\text{Volume flow rate} = \frac{\pi}{4}\left(0.004^2 - 0.002^2\right) \times V_2 = 277 \times 10^{-6} \text{ m}^3$$

or

$$V_2 = 29.4 \text{ m/s}.$$

Now, if the velocity of the jet at the entry side is V_1 m/s, then

$$\frac{V_1}{V_2} = \frac{\left(0.004^2 - 0.002^2\right)}{\left(0.008^2 - 0.004^2\right)}$$

or

$$V_1 = \frac{1.2}{4.8} \times V_2 = 7.35 \text{ m/s}$$

Now applying Bernoulli's theorem for streamline flow:

$$\frac{V_1^2}{2} + \frac{P_1}{\rho} = \frac{V_2^2}{2} + \frac{P_2}{\rho}$$

Here P_2 is atmospheric pressure, that is 1 bar or 10^5 Pa (Pa = N/m²). or

$$\frac{7.35^2}{2} + \frac{P_1}{1000} = \frac{29.4^2}{2} + \frac{10^5}{1000}$$

$$P_1 = 505169 \text{ N/m}^2 = 5.05 \text{ bar}$$

So, the velocity of the water-jet at the exit and pressure at the supply side are 29.4 m/s and 5.05 bar, respectively.

REFERENCES

Adanur, S. 2001. *Handbook of Weaving*. Lancaster, UK: Technomic Publishing Company, Inc.

Adanur, S. and Turel, T. 2004. Effects of air and yarn characteristics in air-jet filling insertion: Part II: Yarn velocity measurements with a profiled reed. *Textile Research Journal*, 74: 657–661.

Banerjee, P. K. 2015. *Principles of Fabric Formation*. Boca Raton, FL: CRC Press.

Luenenschloss, J. and Wahhoud, A. 1984. Investigation into the behaviour of yarns in picking with air-jet systems. *Melliand Textilberichte*, 65: 242.

Marks, R. and Robinson, A. T. C. 1976. *Principles of Weaving*. Manchester, UK: The Textile Institute.

Shyam, A. F., Workshop on Opportunities for Industrial Fabrics Producers, 6th November, 2014, Indian Institute of Technology, Delhi.

Technical literature of Toyota water-jet loom LWT 710, 2015, Toyota Industries Corporation, Japan.

Vangheluwe, L. 1999. *Air-Jet Weft Insertion (Textile Progress)*. Manchester, UK: The Textile Institute.

Wahhoud, L., Weide, T. and Jansen, W. 2004. Air index tester: Manufacturing behavior of rotor yarns on air-jet weaving machines. *Melliand International*, 10: 277–279.

CHAPTER **9**

Beat-Up

9.1 OBJECTIVES

The objectives of beat-up motion are to push the newly inserted pickup to the cloth fell and to ensure uniform pick spacing in the fabric. In general, beat-up is done after the insertion of every pick.

9.2 SLEY MOTION

Beat-up is done by the reed which is carried by the sley. The sley derives its rectilinear reciprocating motion from the rotating crankshaft through the connections of crank and crank arm which makes a four-bar linkage mechanism. This is illustrated in Figure 9.1. When the sley moves backwards, that is away from the cloth fell, the space between the top and bottom shed lines in front of the reed increases. Thus shuttle gets sufficient space to enter the shed. When the sley moves forward, it pushes the newly inserted pick to the desired position.

9.2.1 Sley Displacement, Velocity and Acceleration

Let us assume that the radius of the crank and length of the crank arm are r and l, respectively. A schematic diagram of sley movement is shown in Figure 9.2. The crankshaft is rotating in the clockwise direction according to this figure. The angular position of the crank is θ in Figure 9.2. The crank (OA), crank arm (AB), sley (BF) and invisible link (FO) form the four bar linkage here.

From Figure 9.2, radius of the crank = OA = r.

Length of the crank arm = AB = l.

When the crank was at the front centre (0°), the sley was at its forward most position, that is C.

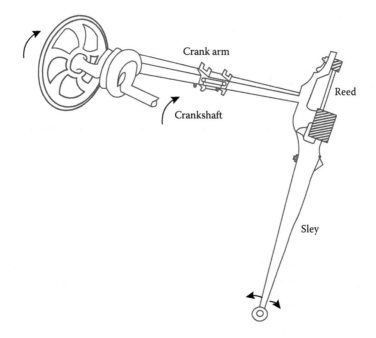

FIGURE 9.1 Sley motion.

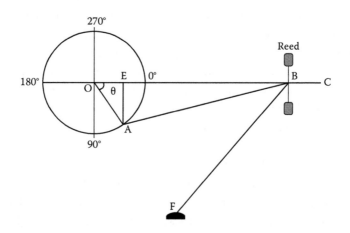

FIGURE 9.2 Schematic diagram of sley motion.

After some time, the crankshaft has moved through an angle θ.

$$\angle EOA = \theta, \quad AE = r\sin\theta, \quad OE = r\cos\theta, \quad OC = r + l$$

So

$$BE = \sqrt{AB^2 - AE^2} = \sqrt{l^2 - r^2\sin^2\theta}$$

or

$$BE = l\left(1 - \frac{r^2 \sin^2 \theta}{l^2}\right)^{\frac{1}{2}}$$

At $\theta°$ revolution of the crankshaft, the sley displacement (measured from point C), $x_\theta = BC$

Now

$$BC = OC - OB = (r+l) - (OE + BE)$$

$$= r + l - \left[r\cos\theta - l\left(1 - \frac{r^2 \sin^2 \theta}{l^2}\right)^{\frac{1}{2}}\right]$$

$$= r(1 - \cos\theta) + l\left[1 - \left(1 - \frac{r^2 \sin^2 \theta}{l^2}\right)^{\frac{1}{2}}\right] \qquad (9.1)$$

As l is much greater than r

$$\frac{r^2 \sin^2 \theta}{l^2} \ll 1$$

So

$$x_\theta = r(1 - \cos\theta) + l\left[1 - \left\{1 - \frac{1}{2} \times \frac{r^2 \sin^2 \theta}{l^2} + \frac{\frac{1}{2}\left(\frac{1}{2}-1\right)}{2} \times \left(\frac{r^2 \sin^2 \theta}{l^2}\right)^2 \cdots\right\}\right]$$

or

$$x_\theta = r(1 - \cos\theta)$$

$$+ l\left[1 - \left\{1 - \frac{1}{2} \times \frac{r^2 \sin^2 \theta}{l^2}\right\}\right] \quad \text{(Neglecting terms which are } \ll 1\text{)}$$

or

$$x_\theta = r(1 - \cos\theta) + l\left[\frac{1}{2} \times \frac{r^2 \sin^2\theta}{l^2}\right]$$

or

$$x_\theta = r\left(1 - \cos\theta + \frac{r\sin^2\theta}{2l}\right) \qquad (9.2)$$

At $\theta = 0°$, displacement $x_0 = 0$

At $\theta = 90°$, displacement $x_{90} = r\left(1 + \frac{r}{2l}\right)$

At $\theta = 180°$, displacement $x_{180} = 2r$

At $\theta = 270°$, displacement $x_{270} = x_\theta = r\left(1 + \frac{r}{2l}\right)$

It can be easily understood from these expressions that during its backward movement (0°–180°), the sley has more displacement from 0° to 90° rotation of the crankshaft than from 90° to 180° rotation of the crankshaft. Similarly during its forward movement (180°–360°), the sley has less displacement from 180° to 270° rotation of the crankshaft than from 270° to 360° rotation of the crankshaft. It will be discussed in more detail subsequently.

Pure simple harmonic motion (SHM) would give the displacement equation as

$$x_{\theta(SHM)} = r(1 - \cos\theta) \qquad (9.3)$$

Therefore, Equation 9.2 indicates that sley motion deviates from SHM as an extra term $r\sin^2\theta/2l$ is added. If the length of the crank arm is much greater than the radius of crank, that is $l \gg r$, then the extra term becomes close to zero and the sley movement almost follows SHM.

$$\text{Sley velocity} = v_\theta = \frac{dx_\theta}{dt} = \frac{dx_\theta}{d\theta} \cdot \frac{d\theta}{dt} = \frac{dx_\theta}{d\theta} \cdot \omega \qquad (9.4)$$

where ω is the angular velocity of the crankshaft.

$$v_\theta = \omega r \left(\sin\theta + \frac{r\sin 2\theta}{2l} \right) \tag{9.5}$$

$$\text{Sley acceleration} = f_\theta = \frac{dv_\theta}{dt} = \frac{dv_\theta}{d\theta} \cdot \frac{d\theta}{dt} = \frac{dv_\theta}{d\theta} \cdot \omega$$

or

$$f = \omega^2 r \left(\cos\theta + \frac{r\cos 2\theta}{l} \right) \tag{9.6}$$

At $\theta = 0°$, $\cos\theta = 1$ and $\cos 2\theta = 1$, therefore $f_0 = \omega^2 r \left(1 + \frac{r}{l} \right)$

At $\theta = 180°$, $\cos\theta = -1$ and $\cos 2\theta = 1$, $f_{180} = -\omega^2 r \left(1 - \frac{r}{l} \right)$

Here f_0 and f_{180} are the maximum and the minimum values of sley acceleration, respectively. The minus sign indicates the acceleration in the opposite direction. Besides, the magnitude of acceleration is also not same for the 0° and 180° angular positions of crank (Booth, 1977). The absolute value of acceleration is higher at 0° angular position of crank. The plots of sley displacement (using Equation 9.2), velocity (using Equation 9.5) and acceleration (using Equation 9.6) are shown in Figures 9.3 through 9.5, respectively, considering the radius of crank as 5 cm. The curve for sley motion is shown by broken line, whereas the solid line indicates the curve for SHM. From Figure 9.3, it can be inferred that the sley attains half of its maximum displacement, that is 5 cm in this case, earlier during its backward movement (from 0° to 180°) as compared to SHM. From Figure 9.4, it can be inferred that the sley attains its maximum velocity earlier as compared to SHM during its backward movement (from 0° to 180°).

Sley displacement curves using Equation 9.2 for various combinations of r and l are shown in Figure 9.6. The radius of crank has been considered as 5 cm here. It is noted that as the ratio r/l increases, the shape of the displacement curve changes. For an infinitely long crank arm, sley motion becomes SHM as r/l tends to 0.

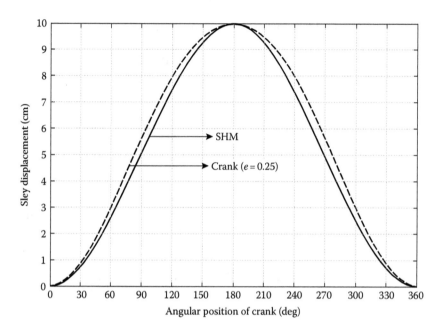

FIGURE 9.3   Sley displacement *vs* crankshaft rotation.

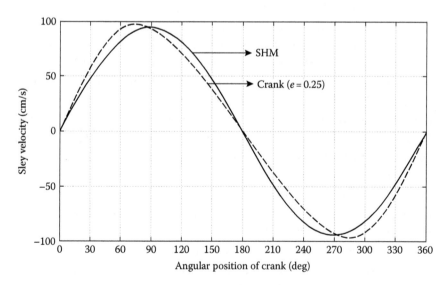

FIGURE 9.4   Sley velocity *vs* crankshaft rotation.

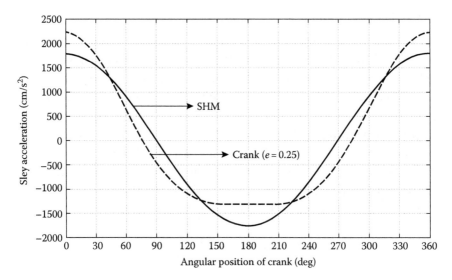

FIGURE 9.5   Sley acceleration *vs* crankshaft rotation.

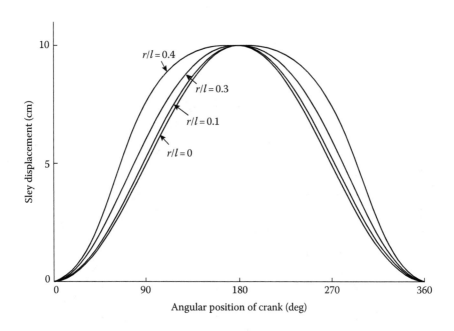

FIGURE 9.6   Sley displacement *vs* crankshaft rotation at different ratios of *r* and *l*.

9.2.2 Sley Eccentricity

As sley motion deviates from SHM, during its backward journey, the sley has more displacement from 0° to 90° rotation of the crankshaft than from 90° to 180° rotation of the crankshaft. Similarly during its forward journey, the sley has less displacement from 180° to 270° rotation of the crankshaft than from 270° to 360° rotation of the crankshaft. This difference in the sley displacement during its backward and forward movement is termed sley eccentricity (e). In the case of SHM, the displacement is same from 0° to 90°, 90° to 180°, 180° to 270° and 270° to 360°. The sley displacement is shown by the broken line in Figure 9.7. The solid line shows the displacement for SHM.

Let the sley displacement from $\theta = 0°$ to $\theta = 90°$ be X_a.

$$X_a = x_{90} - x_0 = r\left(1+\frac{r}{2l}\right) - 0$$

$$= r\left(1+\frac{r}{2l}\right) \tag{9.7}$$

Sley displacement from $\theta = 90°$ to $\theta = 180°$ is X_b.

$$X_b = x_{180} - x_{90} = 2r - r\left(1+\frac{r}{2l}\right)$$

$$= r\left(1-\frac{r}{2l}\right) \tag{9.8}$$

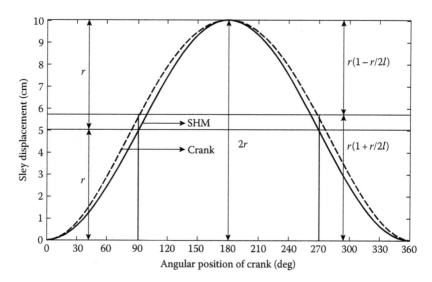

FIGURE 9.7 Sley displacement *vs* angular position of crank.

Similarly, sley displacement from $\theta = 180°$ to $\theta = 270°$ is given by

$$X_b = r\left(1 - \frac{r}{2l}\right)$$

and sley displacement from $\theta = 270°$ to $\theta = 360°$ is given by

$$X_a = r\left(1 + \frac{r}{2l}\right)$$

So, by definition

$$e = X_a - X_b$$

or

$$e = r\left(1 + \frac{r}{2l}\right) - r\left(1 - \frac{r}{2l}\right)$$

or

$$e = \frac{r^2}{l} \qquad (9.9)$$

When the sley displacement is plotted using a displacement scale relative to crank radius (r), the deviation of sley displacement as compared to that of SHM is determined by the ratio r/l. This is commonly considered as an indicator of sley eccentricity.

9.2.2.1 Calculation Related to Sley Eccentricity

It has been mentioned earlier that, the sley attains the following two things earlier, as compared to SHM, during its backward movement (0°–180°) and vice versa.

1. Half of its maximum displacement

2. Maximum velocity

Let us assume that the sley eccentricity ($e = r/l$) value is $= 0.1$

$$\text{Displacement} = s = r\left(1 - \cos\theta + \frac{r\sin^2\theta}{2l}\right)$$

Here

$$\text{Half of the maximum displacement} = s = 2r \times \frac{1}{2} = r.$$

Now

$$e = \frac{r}{l} = 0.1$$

So

$$r = r\left(1 - \cos\theta + 0.1 \times \frac{\sin^2\theta}{2}\right)$$

or

$$\sin^2\theta - 20\cos\theta = 0$$

or

$$\cos^2\theta + 20\cos\theta - 1 = 0$$

Hence

$$\cos\theta = \frac{-20 \pm \sqrt{400 + 4}}{2} = 0.05$$

or

$$\theta \approx 87° \text{ and } 273°$$

So the span during which the sley displacement is greater than half of maximum displacement = 273° − 87° = 186°.

The calculation can be extended for different values of sley eccentricity (see Table 9.1).

So, as the sley eccentricity increases, the sley remains towards the back side for a longer duration, thus providing more time for the uninterrupted shuttle flight.

TABLE 9.1   Effect of Sley Eccentricity

Eccentricity Value	Angular Position of the Crankshaft at Half of Maximum Displacement	Period during Which the Sley Displacement Is Greater Than Half of Maximum Displacement	Angular position of the Crankshaft for Maximum Sley Velocity
0.0 (SHM)	90 and 270	180	90 and 270
0.1	87 and 273	186	84 and 276
0.2	83 and 277	194	79 and 281
0.3	80 and 280	200	75 and 285
0.5	75 and 285	210	68.5 and 291.5

9.2.2.2 Effect of Sley Eccentricity

It is shown in the following section that the force required to drive the sley is proportional to the equivalent mass of the sley and square of the loom speed. It is also dependent on the sley eccentricity ratio (*e*). Obviously, increasing the sley eccentricity tends to increase the effectiveness of beat-up by producing high beat-up force. This is suitable for weaving heavy fabrics made from coarse yarns which require a high beat-up force. However, a high sley eccentricity increases the force acting on the crankpins, sword-pins, cranks, crank arms, crankshaft and their bearing and indirectly on the loom frame. Therefore, a high value of sley eccentricity demands more robust loom parts in order to prevent excessive wear and vibration (Marks and Robinson, 1976).

With SHM ($e = 0$), the sley attains its maximum velocity and exactly half of its maximum displacement at 90° and again at 270°. As *e* increases, the sley attains half of its maximum displacement earlier during its backward movement and later during its forward movement. This phenomenon is depicted in Figure 9.8. The different sley eccentricities have been created by changing the length of the crank arm. The radius of the crank has been considered to be constant (5 cm). As the sley eccentricity increases, the sley remains longer nearer to its most backward position and hence more time will be available for the passage of shuttle through the shed ($\theta_2 > \theta_1 > \theta$). This will be advantageous for wide width loom.

9.3 FORCE, TORQUE AND POWER REQUIRED TO DRIVE THE SLEY

Figure 9.9 depicts the forces acting during the sley motion. The movement of the point S and thus the acceleration and deceleration of reed and raceboard take place along the horizontal line. The force required for this (*F*) can be calculated by multiplying the equivalent mass of sley and its acceleration.

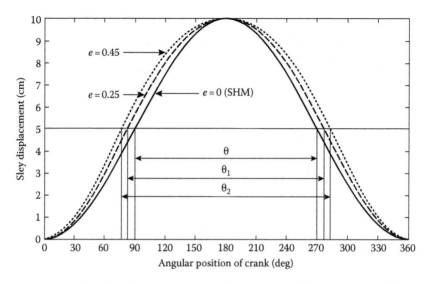

FIGURE 9.8 Sley displacement *vs* angular position of crank with different eccentricity.

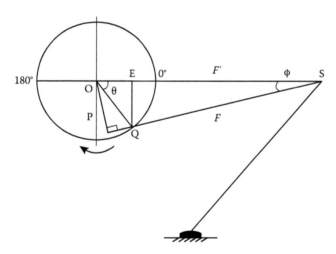

FIGURE 9.9 Forces acting during sley motion.

However, the force (F') that creates rotational torque about the axis of the crankshaft acts along the crank arm (SQ).

From Figure 9.9:

$$\angle QOS = \theta, \quad \angle QSO = \phi, \quad \angle OPQ = 90°, \quad OQ = r$$

$$\angle OQP = (\theta + \phi), \quad OP = r\sin(\theta + \phi)$$

Total reaction force acting on sley $= F = Mf$

where

M is the equivalent mass of the sley and associated parts
f is the acceleration of sley

Now from Equation 9.6

$$f = \omega^2 r \left(\cos\theta + \frac{r}{l}\cos 2\theta \right)$$

$$F = M\omega^2 r \left(\cos\theta + \frac{r}{l}\cos 2\theta \right) \tag{9.10}$$

Force acting on the crank arm = $F' = F \sec\phi$

Therefore, torque (τ) about point O is $F' \times OP$
Or
$\tau = F \sec\phi\, r \sin(\theta + \phi)$
Assuming ϕ is too small in comparison to θ,

$$\tau = Fr\sin\theta$$

$$= M\omega^2 r^2 \sin\theta \left(\cos\theta + \frac{r}{l}\cos 2\theta \right)$$

$$= \frac{M\omega^2 r^2}{2}\left\{ \sin 2\theta + \frac{r}{l}(2\cos 2\theta \sin\theta) \right\}$$

or

$$\tau = \frac{M\omega^2 r^2}{2}\left\{ \sin 2\theta + \frac{r}{l}(\sin 3\theta - \sin\theta) \right\} \tag{9.11}$$

Therefore, torque required is proportional to the equivalent mass of the sley and the square of the loom speed and crank radius (Lord and Mohamed, 1982). The torque acting on sley during its motion as a function of angular position of crankshaft is shown in Figure 9.10.

Now

$$\text{Power needed to drive the sley} = P = \tau\omega$$

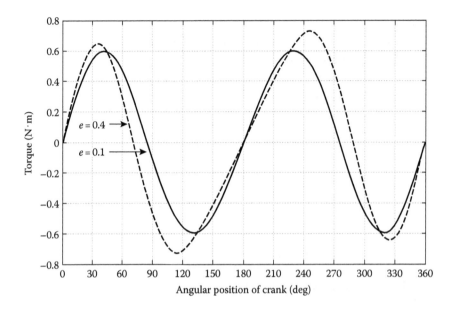

FIGURE 9.10 Sley torque *vs* angular position of crank for different eccentricities.

or

$$P = \frac{M\omega^3 r^2}{2}\left\{\sin 2\theta + \frac{r}{l}\left(\sin 3\theta - \sin\theta\right)\right\} \qquad (9.12)$$

From Equation 9.12, it can be inferred that the power required is proportional to the equivalent mass of the sley, the cube of the loom speed and the square of the crank radius.

9.4 ANALYSIS OF MOTIONS OF VARIOUS POINTS ON THE SLEY

Sley geometry and coordinates of crank and crank arm are shown in Figures 9.11 and 9.12, respectively. P is any point on the axis of crank arm. The motions of points A, B, E and P during sley movement are different, and can be understood with simple analysis.

In Figure 9.12, crank radius = OA = *r* and length of crank arm = AB = *l*.

The coordinates of point A = (*r* cos θ, *r* sin θ) and locus of the point A describes a circle.

Point E, which is the projection of crankpin on the *x*-axis, performs SHM. The motion of the point B is more complicated and deviates from SHM. The motion of point B can be termed eccentric motion which will be dependent on the sley eccentricity.

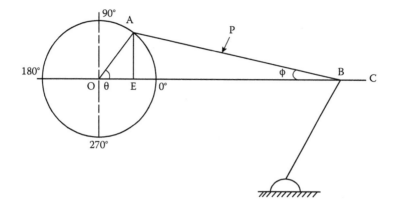

FIGURE 9.11 Sley geometry.

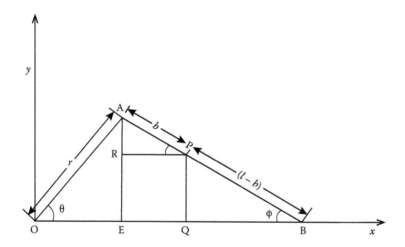

FIGURE 9.12 Coordinates of crank and crank arm.

Now

$$AE = r \sin \theta = l \sin \phi \text{ and } AR = b \sin \phi$$

$$RE = AE - AR = (l - b)\sin \phi$$

$$= (l - b)\frac{r}{l}\sin \theta$$

For any point P on the axis of the crank arm at a distance b from the point A,

$$y = PQ = RE = \frac{l-b}{l}r \sin \theta \tag{9.13}$$

and

$$x = OQ = OE + EQ$$

or

$$x = OE + RP$$

or

$$x = r\cos\theta + b\cos\varphi$$

or

$$x = r\cos\theta + b\sqrt{1 - \sin^2\phi}$$

or

$$x = r\cos\theta + b\sqrt{1 - \left(\frac{r}{l}\right)^2 \sin^2\theta} \tag{9.14}$$

Using Equations 9.13 and 9.14, for the special case where $r = l$

$$x = (r + b)\cos\theta \tag{9.15}$$

$$y = (r - b)\sin\theta \tag{9.16}$$

Now

$$\sin^2\theta + \cos^2\theta = 1$$

or

$$\left(\frac{x}{r+b}\right)^2 + \left(\frac{y}{r-b}\right)^2 = 1 \tag{9.17}$$

From Equation 9.17, it can be inferred that in this particular case where $r = l$, the locus each point on the axis of the crank arm describes an ellipse.

Therefore, motion of point A is circular, motion of point B is eccentric, motion of point E is SHM and motion of point P is elliptic for a special case where $r = l$.

9.5 WEAVING RESISTANCE

During beat-up, the reed experiences resistance due to the following reasons:

- The reed needs to overcome friction between the pick and warp.

- As the reed pushes the pick towards the cloth fell, the warp yarns are bent around the pick. Both the warp and the pick attain crimped configuration.

- The cloth fell is displaced by the reed.

The force that the reed exerts on the cloth fell at the time of beat-up is termed beat-up force. Weaving resistance is the equal and opposite reaction force generated in response to beat-up force. When the reed is away from the cloth fell, the tension in the warp and tension in the fabric is the same. However, they act in opposite directions and thus the cloth fell remains stationary. During the beat-up, the reed displaces the cloth fell. Thus the warp gets stretched and the fabric gets contracted. As a result the tension in the warp increases and tension in the fabric decreases. This difference between instantaneous warp tension and fabric tension is balanced by the beat-up force. The situation can be understood from the spring model of fabric beat-up as shown in Figure 9.13.

Z is the distance by which the cloth fell is displaced during beat-up from the position it occupied before beat-up.

L_w, L_f are the free length of warp and fabric, respectively, as shown in Figure 9.14.

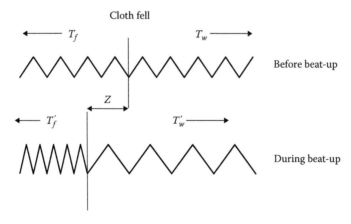

FIGURE 9.13   Spring model of beat-up.

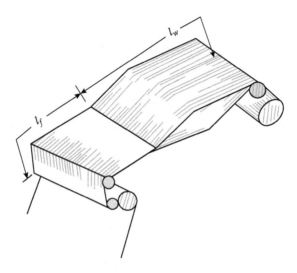

FIGURE 9.14 Free length of warp and fabric.

R is the weaving resistance.

T_0 is the basic tension in warp and fabric just before beat-up.

T_w, T_f are the tension in warp and fabric, respectively, just before beat-up.

T_w', T_f' are the tension in warp and fabric, respectively, at the time of beat-up.

E_w, E_f are the elastic modulus of warp and fabric, respectively.

Here R, T_0, T_w, T_f, T_w' and T_f' are expressed in terms of tension per end. Now just before beat-up, $T_w = T_f = T_0$.

At the time of beat-up, warp tension $= T_w' = T_w + \Delta T_w$ and fabric tension $= T_f' = T_f - \Delta T_f$.

Weaving resistance is the difference between warp tension and fabric tension during beat-up.

Therefore

$$R = T_w' - T_f'$$

or

$$R = \left(T_w + \Delta T_w\right) - \left(T_f - \Delta T_f\right)$$

or

$$R = \left(\Delta T_w - \Delta T_f\right) \text{ as } T_w = T_f = T_0$$

Now at the time of beat-up, warp is extended and fabric is contracted by the length Z.

Thus, strain in warp $= Z/L_w$ and strain in fabric $= Z/L_f$.

So

$$\Delta T_w = \frac{E_w Z}{L_w} \quad \text{and} \quad \Delta T_f = \frac{E_f Z}{L_f}$$

Therefore

$$R = \frac{E_w Z}{L_w} + \frac{E_f Z}{L_f} = Z\left(\frac{E_w}{L_w} + \frac{E_f}{L_f}\right) \tag{9.18}$$

From Equation 9.18, it is noted that the weaving resistance is independent of basic warp tension (T_w or T_0) just before beat-up. Weaving resistance increases with the increase in the displacement of the cloth fell (Z) during beat-up.

9.5.1 Cloth Fell Position and Pick Spacing

The reed reaches the most forward position of its sweep when the crank is at the front centre (0°). This position of the reed can be considered as a reference line as shown in Figure 9.15. The position of the cloth fell before beat-up can be at the front or behind this reference line. Let the distance between the reference line and the cloth fell position before beat-up be L. This distance is considered to be negative as the cloth fell is positioned behind the reference line and vice versa. During beat-up, the distance between the new pick and cloth fell is reduced to the pick spacing (S) as the newly inserted pick occupies a position in front of the reed. The cloth fell occupies a position at a distance S in the front of reference line.

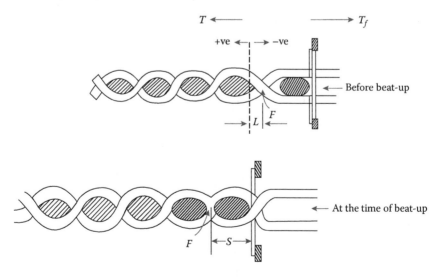

FIGURE 9.15 Cloth fell position before and after beat-up.

Since the cloth fell was at a distance L before the beat-up and a distance S at the time of beat-up, the distance by which the cloth fell has been displaced is given by

$$Z = (S - L) \tag{9.19}$$

From Equations 9.18 and 9.19, we get

$$R = (S - L)\left(\frac{E_w}{L_w} + \frac{E_f}{L_f}\right) \tag{9.20}$$

Empirically, the relation between the pick spacing S and weaving resistance R corresponding to that pick spacing is given by

$$R = \frac{K}{S - S_{min}} \tag{9.21}$$

where
  K is an empirical constant
  S_{min} is the minimum pick spacing which is imagined to be obtained when weaving resistance approaches to infinity

Therefore, as $S \rightarrow S_{min}$, $R \rightarrow \infty$
From Equations 9.20 and 9.21, we get

$$\frac{K}{S - S_{min}} = (S - L)\left(\frac{E_w}{L_w} + \frac{E_f}{L_f}\right)$$

or

$$(S - L) = \frac{K}{S - S_{min}}\left(\frac{E_w}{L_w} + \frac{E_f}{L_f}\right)^{-1}$$

or

$$(S - L) = \frac{K'}{S - S_{min}}$$

where

$$K' = K\left(\frac{E_w}{L_w} + \frac{E_f}{L_f}\right)^{-1}$$

Therefore

$$L = S - \frac{K'}{S - S_{min}} \qquad (9.22)$$

From Equation 9.22, one can comprehend the position that the cloth fell should occupy (L) prior to beat-up in order to produce the desired pick spacing (S) at subsequent beat-up (Greenwood, 1975). From Equation 9.21, if the pick spacing (S) is very high, then ($S - S_{min}$) becomes very high and weaving resistance R becomes negligible. So $L = S$ as ($S - L$) = 0 (Equation 9.20). On the other hand, when the pick spacing (S) is very low, ($S - S_{min}$) becomes very low. Then the following expression can be written by neglecting S of Equation 9.22 (Greenwood, 1975).

$$L = - \frac{K'}{S - S_{min}} \qquad (9.23)$$

Now, putting the value of the desired pick spacing in Equation 9.23, the position of cloth fell before the beat-up can be found. Here, the negative value implies that the cloth fell occupies a position behind the reference line (most forward position of reed) before beat-up. The relationship between cloth fell position and pick spacing is shown in Figure 9.16 for a particular combination of warp, weft and weave.

9.5.2 Bumping

Figure 9.17 shows the tension variation in warp (solid line) and fabric (broken line) during the weaving cycle. Barring the instance of beat-up, the tension in the warp and fabric follows similar pattern. However, at the time of beat-up, warp tension increases, whereas fabric tension decreases. If fabric tension during beat-up becomes zero, that is $T'_f = 0$, then the fabric becomes momentarily slack (Figure 9.17). This condition is termed bumping. It is easily detectable by a noise when the cloth becomes suddenly taut as the sley moves backward after the beat-up. Under bumping condition, absence of fabric tension at the time of beat-up reduces weaving resistance and hence the effectiveness of beat-up. Bumping can be avoided by increasing the basic warp and fabric tension just before beat-up (T_0). However, this may be detrimental from warp breakage rate and loom efficiency viewpoint. Figure 9.18 illustrates the normal condition during weaving without any bumping.

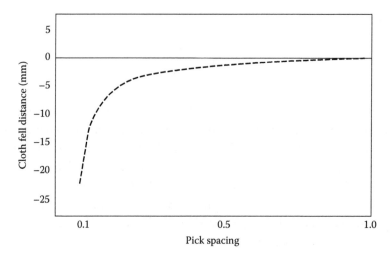

FIGURE 9.16 Relationship between cloth fell position and pick spacing.

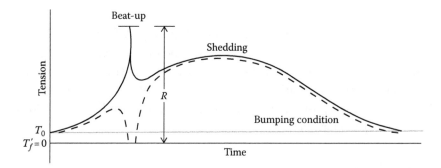

FIGURE 9.17 Bumping condition during beat-up.

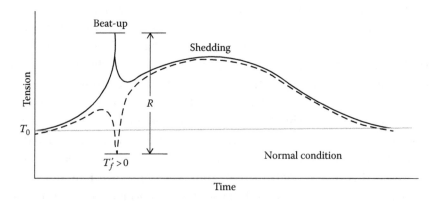

FIGURE 9.18 Normal condition during beat-up.

9.6 FACTORS INFLUENCING THE BEAT-UP FORCE

The following factors influence the beat-up force during weaving.

- Warp tension

- Pick spacing

- Weft yarn linear density

- The timing of shedding

Bullerwell and Mohamed (1991) and Shih et al. (1995) have demonstrated that higher warp tension leads to higher beat-up force as shown in Figure 9.19. The beat-up force starts to generate as the reed commences to push the newly inserted pick towards the cloth fell by overcoming the bending and frictional resistances offered by the warp. The beat-up takes place over a certain period of time, while at 0° the beat-up force reaches its peak as cloth fell displacement is the maximum at this moment.

Lower pick spacing (i.e. higher picks per cm) requires higher beat-up force as higher cloth fell displacement is needed in this case. The beat-up force increases further if the weft yarns are coarser. Therefore, lower pick spacing and coarser weft yarn lead to drastic rise in the beat-up force as shown by the broken line of Figure 9.20. The time span (pulse), during which the beat-up force acts on the warp yarns, also increases with lower pick spacing and coarser weft yarns. Three lines in Figure 9.21 depict the time span of beat-up force for three combinations of pick spacing and weft yarn linear density (lower pick spacing and higher linear density, higher pick spacing and higher linear density, higher pick spacing and lower

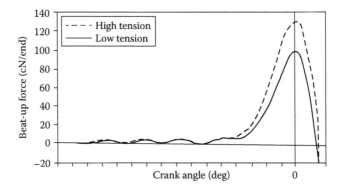

FIGURE 9.19   Influence of warp tension on beat-up force.

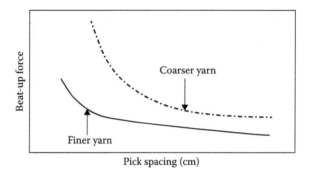

FIGURE 9.20 Influence of pick spacing and weft yarn linear density on beat-up force.

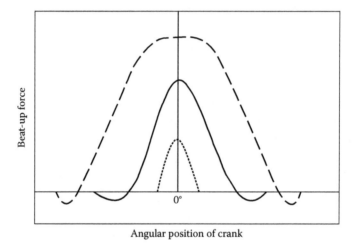

FIGURE 9.21 Influence of pick spacing and weft yarn linear density on beat-up force duration.

linear density). The first combination (lower pick spacing and higher linear density, i.e. coarser yarn) yields the largest span as well as the highest beat-up force as shown by the broken line. On the other hand, the last combination (higher pick spacing and lower linear density, i.e. finer yarn) produces the smallest span as well as the lowest beat-up force as shown by the dotted line. The solid line represents the situation for higher pick spacing and higher weft linear density.

If the shed closes or levels early, then beat-up takes place at crossed shed. Thus the reed has to overcome additional bending resistance of warp yarns. Thus the beat-up force increases as compared to that of beat-up at open shed.

9.7 TEMPLE

During beat-up, the reed pushes the pick against the frictional and bending resistances of warp yarns. As the interlacement takes place between ends and pick, the latter contracts along its length. So the width of the fabric becomes less than that of the warp sheet. If it is allowed to happen freely, then the ends will form a wedge shape near the cloth fell as shown on the right-hand side of Figure 9.22. This will lead to severe abrasion between the warp and the reed during the movement of the latter. This will cause large number of end breaks, thus making weaving very difficult. The problem is prevented by using temple which is a device to hold the fabric in desired width near the cloth fell. Two sides of the fabric, near the selvedge, are gripped by the temples as shown on the right-hand side of Figure 9.22. Temples are of the following types:

- Ring temple (ring with spikes)

- Rubber temples (without spikes and used for delicate fabrics)

- Steel roller temples (no ring but spikes are present)

- Full-width temple

Figure 9.23 shows a ring temple mounted on the right-hand side of a loom. Generally two temples, positioned at the two sides of the loom, cover partial width of fabric. Full-width temple is used for weaving very heavy fabrics having very high picks per cm which require very high beat-up force.

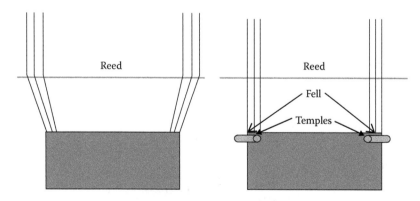

FIGURE 9.22   Situations during weaving without and with temple.

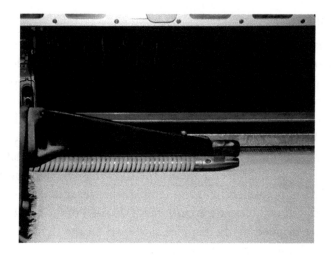

FIGURE 9.23 Ring temple.

NUMERICAL PROBLEMS

9.1 The angular position of the crank is represented by angle θ. When the acceleration of sley is zero (or velocity is maximum), then show the following:

$$\theta = \cos^{-1} \frac{l\left(\sqrt{1+8\left(\dfrac{r}{l}\right)^2}-1\right)}{4r}$$

Solution:

Acceleration of sley $= a = r\omega^2 (\cos\theta + r/l\cos 2\theta)$

$$= r\omega^2 \left[\cos\theta + \frac{r}{l}\left(2\cos^2\theta - 1\right)\right]$$

For acceleration to be zero

$$a = 0$$

So

$$2r\cos^2\theta + l\cos\theta - r = 0$$

or

$$\cos\theta = \frac{-l \pm \sqrt{l^2 - 4 \times 2r \times -r}}{2 \times 2r}$$

or

$$\theta = \cos^{-1} \frac{-l \pm l\sqrt{1 + 8\left(\dfrac{r}{l}\right)^2}}{4r}$$

Considering the positive sign only,

$$\theta = \cos^{-1} \frac{l\left(\sqrt{1 + 8\left(\dfrac{r}{l}\right)^2} - 1\right)}{4r}$$

$$= \cos^{-1} \frac{\sqrt{1 + 8e^2} - 1}{4e}; \quad \text{as} \left(\frac{r}{l} = e\right)$$

9.2 Draw the velocity and acceleration curves for a sley, having eccentricity of 0.5, against the angular position of the crank. Compare them with those of SHM.

Solution:
Here

$$\text{Sley eccentricity } e = r/l = 0.5$$

Velocity for sley is given by

$$v_e = r\omega \left(\sin\theta + \frac{e\sin 2\theta}{2}\right)$$

Velocity for SHM is given by

$$v_{SHM} = r\omega \sin\theta$$

$\sin 2\theta = 0$
when $\theta = 90°, 180°, 270°, 360°$
So v_e and v_{SHM} are equal at the aforementioned values of θ.
For, maximum velocity of sley, acceleration will be zero.
It has been shown that

$$\theta_{V\max} = \cos^{-1} \frac{\sqrt{1 + 8e^2} - 1}{4e} = \cos^{-1} \frac{\sqrt{1 + 2} - 1}{2} \quad (\text{as } e = 0.5)$$

$$= \cos^{-1} 0.366 = 68.5°, 291.5°$$

So at 68.5° and 291.5° angular positions of crank, maximum sley velocity will be attained. Now, putting $\theta = 68.5°$ and 291.5° in the expression of sley velocity, the value of maximum velocity can be obtained in terms of $r\omega$. This will be 1.1 $r\omega$.

When θ is less than 90°, the term $(e\sin 2\theta)/2$ is positive and therefore the velocity of sley will be more than that of SHM. However, when θ is greater than 90°, the term $(e\sin 2\theta)/2$ is negative, and therefore, the velocity for sley will be less than that of SHM. The velocity curves for sley and SHM will intersect each other at 90°, 180° and 270° as shown in Figure 9.24.

$$\text{Acceleration of sley} = f_e = r\omega^2(\cos\theta + e\cos 2\theta)$$

$$\text{Acceleration in SHM} = f_{SHM} = r\omega^2(\cos\theta)$$

When $\theta = 0°$,

$$f_e(\theta = 0°) = r\omega^2(1+e) = 1.5r\omega^2$$

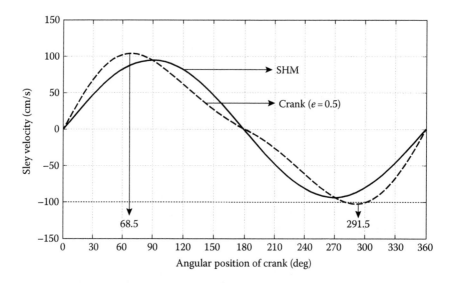

FIGURE 9.24 Velocity profile for sley eccentricity of 0.5.

When $\theta = 180°$,

$$f_e(\theta = 180°) = r\omega^2(-1+e) = -0.5r\omega^2$$

Now differentiating the expression of acceleration with respect to time,

$$\frac{df_e}{dt} = r\omega^3(-\sin\theta - 2e\sin 2\theta)$$

$$= -r\omega^3(\sin\theta + \sin 2\theta)$$

If $da_e/dt = 0$ then

$$(\sin\theta + \sin 2\theta) = 0$$

or

$$\sin\theta(1 + 2\cos\theta) = 0$$

So

$$\sin\theta = 0 \quad \text{or} \quad \cos\theta = -0.5$$

Therefore

$$\theta = 0°, 180°, 120° \text{ and } 240°$$

Therefore, at these four values of θ, the slope of the acceleration curve will be zero. Using these values of θ in the expression of acceleration of sley, the values of acceleration can be calculated in terms of $r\omega^2$.
When $\theta = 120°$ or $240°$,

$$f_e(\theta = 120°, 240°) = -0.75r\omega^2$$

Now, using this information, the acceleration curve of the sley can be plotted as shown in Figure 9.25. As the maximum velocity is attained when $\theta = 68.5°$ and $291.5°$, acceleration at these two points will be zero.

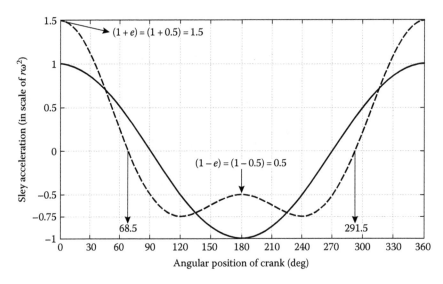

FIGURE 9.25 Acceleration profile for sley eccentricity of 0.5.

9.3 A loom is running at 240 picks per minute. The shuttle can enter and remain within the shed when the sley displacement is at least 50% of its maximum displacement. If the loom width is 1.75 m and shuttle length is 0.25 m, then determine the minimum average velocity of the shuttle (m/s) during its flight assuming $e = 1/3$.

Solution:

$$\text{Sley displacement} = s = r\left(1 - \cos\theta + \frac{r\sin^2\theta}{2l}\right)$$

Maximum sley displacement = 2r, where r is the radius of crank.
   Here

$$s = 2r \times \frac{1}{2} = r \quad \text{and} \quad e = \frac{r}{l} = \frac{1}{3}$$

The shuttle can stay in the shed when the sley displacement is at least r.
   So

$$r = r\left(1 - \cos\theta + \frac{\sin^2\theta}{6}\right)$$

   or

$$\sin^2\theta - 6\cos\theta = 0$$

or

$$\cos^2\theta + 6\cos\theta - 1 = 0$$

Hence

$$\cos\theta = \frac{-6 \pm \sqrt{36 + 4}}{2} = 0.162.$$

or

$$\theta \approx 80° \text{ and } 280°$$

So

The maximum possible span of crankshaft movement for shuttle flight

$$= 280° - 80°$$
$$= 200°$$

Maximum available time for flight $= \dfrac{60}{240} \times \dfrac{200}{360}$ s $= 0.139$ s.

Distance travelled during this time $= (1.75 + 0.25)$ m $= 2$ m.
So

the minimum average velocity of shuttle $= \dfrac{2}{0.139} = 14.4$ m/s.

9.4 The tensile modulus of a woven fabric and corresponding warp sheet is 20 N/cm width and 30 N/cm, respectively. Fabric width is 150 cm. If the cloth fell displacement is 1 cm during beat-up and the free length of fabric and warp sheet is 50 and 100 cm, respectively, then calculate the weaving resistance. Determine the minimum basic tension to prevent bumping.

Solution:
Here
Weaving resistance = R.
Modulus of fabric and warp sheet is E_f and E_w, respectively.
Length of fabric and warp sheet is L_f and L_w, respectively.
Cloth fell displacement during beat-up = Z.

Basic warp tension = T_0.

$$\text{Weaving resistance} = R = Z\left(\frac{E_f}{L_f} + \frac{E_w}{L_w}\right)$$

$$R/\text{cm} = 1\left(\frac{20}{50} + \frac{30}{100}\right)$$

$$= 1(0.4 + 0.3) = 0.7 \text{ N/cm}.$$

So

$$\text{Total weaving resistance} = 0.7 \times 150 = 105 \text{ N}$$

Minimum tension to prevent bumping:

At limiting condition for bumping the fabric tension will be zero during beat-up.

So

$$0 = T_0 - Z\frac{E_f}{L_f}$$

$$= T_0 - 1 \times \frac{20}{50}$$

or

$$T_0 = 0.4 \text{ N/cm}.$$

So

Basic warp tension = 0.4 ×150 = 60 N.

So weaving resistance is 105 N and basic warp tension to prevent bumping is 60 N.

REFERENCES

Booth, J. E. 1977. *Textile Mathematics*, Vol. III. Manchester, UK: The Textile Institute.

Bullerwell, A. C. and Mohamed, M. H. 1991. Measuring beat-up force on a water jet loom. *Textile Research Journal*, 61: 214–222.

Greenwood, K. 1975. *Weaving-Control of Fabric Structure*. Merrow, UK: Merrow Technical Library.

Lord, P. R. and Mohamed, M. H. 1982. *Weaving: Conversion of Yarn to Fabric*, 2nd edn. Merrow, UK: Merrow Technical Library.

Marks, R. and Robinson, A. T. C. 1976. *Principles of Weaving*. Manchester, UK: The Textile Institute.

Shih, Y., Mohamed, M. H., Bullerwell, A. C. and Dao, D. 1995. Analysis of beat-up force during weaving. *Textile Research Journal*, 65: 747–754.

CHAPTER 10

Secondary and Auxiliary Motions

S ECONDARY AND AUXILIARY MOTIONS are needed to operate looms without interruption. The two secondary motions are take-up and let-off. The auxiliary motions are warp stop motion, weft stop motion and warp protector motion.

10.1 TAKE-UP MOTION

10.1.1 Objectives
The objective of take-up is to wind the woven fabric as a new pick is beaten up to the cloth fell in order to maintain the pick spacing constant. For a given loom speed, if the take-up system is operating faster, then picks per cm in the fabric will decrease and vice versa.

10.1.2 Classification
Take-up systems can be classified in three ways. They can be classified as negative or positive based on the driving system. Another classification is intermittent or continuous take-up. Intermittent take-up actuates itself when newly inserted pick is beaten up by the reed. On the other hand, continuous take-up operates continuously to wind the woven fabric. The presence of ratchet and pawl arrangement in the take-up mechanism makes it intermittent type, whereas the presence of worm and worm wheel renders it a continuous one. Take-up system can also be classified as direct or indirect. In the case of the former, the fabric is wound directly on

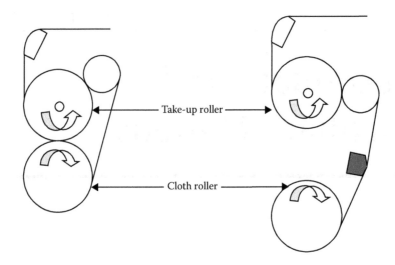

FIGURE 10.1   Indirect fabric take-up systems.

the take-up roller. Therefore, angular velocity of the take-up roller has to be reduced with the increase in effective diameter of take-up roller so that constant pick spacing is achieved. On the other hand, the fabric is wound on the cloth roller and not on the take-up roller in the case of indirect take-up system. The cloth roller can be driven either by frictional contact with the take-up roller or by a friction clutch system. Indirect fabric take-up systems are shown in Figure 10.1.

10.1.2.1 Negative Take-Up

In the case of negative take-up, no positive or direct motion is imparted to the take-up roller to wind the woven fabric. In this system, shown in Figure 10.2, the motion of the rocking shaft actuates a system of levers and a ratchet-pawl mechanism. This is assisted by gravity-aided movement of dead weights which, through a worm and worm-wheel, transmit the rotational motion to take-up roller. Backward motion of the sley (S) or backward swing of the rocking shaft (R) keeps the weights (W) raised through a lever (L_1), keeping the system inactive. The forward motion of the sley sets free the weight system. However, the gear train remains balanced with warp tension. The impulsive blow of sley during beat-up reduces the tension in the woven fabric significantly. Weights (W) going down with gravity force the pawl (P_1) to turn the ratchet by one tooth through lever systems. Gear trains so released wind the newly formed part of the fabric. Retaining pawl (P_2) comes into play to prevent further rotation of gear trains. With little

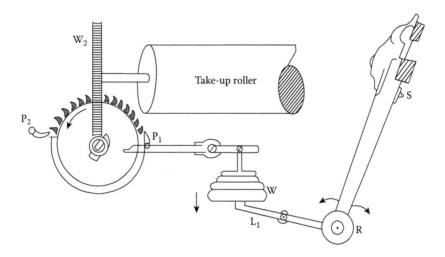

FIGURE 10.2   Negative take-up motion.

control over uniform take-up or pick spacing, this system is suitable only for very coarse fabric like blanket.

10.1.2.2 Positive Take-Up

In positive take-up, motion gets transmitted to take-up roller directly through gear train. The different types of positive take-up are discussed in the following text.

10.1.3 Five-Wheel Take-Up

Five-wheel take-up motion is depicted in Figure 10.3. The typical size (number of teeth) of the gears is also given in the figure. It is a positive intermittent-type take-up motion. CW denotes the number of teeth on change wheel which is in the driver position for five-wheel take-up. The amount of fabric take-up and therefore the rotation required in the take-up roller after every pick are minuscule. This is achieved by keeping the driver wheels smaller than the driven wheels. The ratchet wheel A, having 50 teeth, is turned by one tooth in every pick.

The amount of fabric taken up in each pick, that is pick spacing, can be obtained by calculating the movement of take-up roller for one tooth movement of the ratchet wheel. One tooth movement of ratchet well is equivalent to its 1/50 revolution.

Thus

$$\text{Pick spacing} = 1/50 \times CW/120 \times 15/75 \times 15 = CW/2000 \text{ inch.}$$

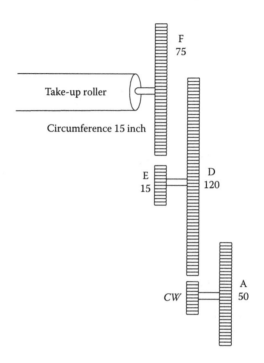

FIGURE 10.3   Five-wheel take-up motion.

Therefore, picks per inch or $PPI = 2000/CW$. So the product of picks per inch and CW is always 2000. For example, in order to achieve 80 PPI, the number of teeth in the change wheel will be $2000/80 = 25$.

10.1.4 Seven-Wheel Take-Up

Seven-wheel take-up motion is shown in Figure 10.4. It is also a positive intermittent-type take-up motion. In contrast to five-wheel take-up, here the change wheel is in driven position. For each pick, the ratchet wheel (A) is turned by one tooth.

The amount of fabric taken up for each pick which corresponds to the pick spacing can be calculated as follows.

Pick spacing = $1/24 \times 36/CW \times 24/89 \times 15/90 \times 15.05 = 1.015/CW$ inch.

Thus

$$\text{Picks per inch or } PPI = CW/1.015$$

or

$$CW = 1.015 \, PPI$$

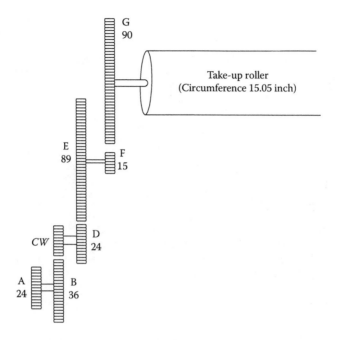

FIGURE 10.4   Seven-wheel take-up motion.

Therefore, the number of teeth in the change wheel is marginally higher (1.5%) than the *PPI* in the fabric in loom state. This 1.5% allowance is given for the lengthwise contraction of fabric. Therefore, the number of teeth in the change wheel becomes almost equal to the *PPI*, when the fabric is taken off the loom. A seven-wheel take-up motion mounted on the loom is shown in Figure 10.5.

Any faulty gear wheel having worn-out or broken tooth or eccentricity in a gear in the train can lead to a periodic variation in pick spacing which produces a fabric defect known as weft bar. If the wavelength (λ) of this periodicity ranges between 1/8 and 10 inch, the effect is readily seen in the fabric. However, if the wavelength is shorter than 1/8 inch or longer than 10 inch, then the human eye cannot identify the periodic defect. Therefore, take-up systems are designed in such a way that the occurrences of such periodicities can be minimised. Calculation of λ for different cases is discussed in the following text.

10.1.4.1 Case I: One Tooth of One Gear Is Faulty

If one tooth on gear G is faulty, that is worn out or broken, then it will create a jerk in the take-up system whenever this faulty tooth meshes with gear F. This will happen only once for one complete revolution of gear G. One revolution of gear G implies 15.05 inch of fabric take-up. Therefore, if one tooth of G is faulty, it produces a periodicity having 15.05 inch

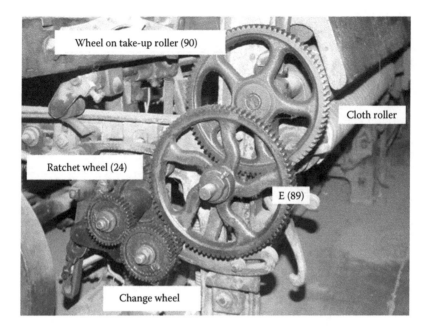

FIGURE 10.5 Seven-wheel take-up motion mounted on a loom.

wavelength. Therefore, the fault will reappear after every 15.05 inch in the fabric. However, the fault will be momentary and will be spread over the fabric corresponding to fabric take-up due to one tooth movement of gear G. This can be considered as the width of the fault. Here the width of the fault will be = 1/90 × 15.05 inch = 0.167 inch. Similarly, if one tooth on gear F is faulty, then it will create a jerk in the take-up system whenever this faulty tooth meshes with gear G. This will happen only once for one complete revolution of gear F. One revolution of gear *F* implies 15/90 × 15.05 = 2.5 inch of fabric take-up. Thus one faulty tooth of gear F produces a periodicity having 2.5 inch wavelength. Now, the width of fault will be the same with that of fault produced by a faulty tooth of gear G. Because, irrespective of the presence of faulty tooth either on gear G or on gear F, the take-up system will experience the same jerk. As the motion is being transmitted to the take-up roller through gear G, the width of the fault will be equivalent to fabric take-up due to one tooth movement of gear G. So the width of the fault will be = 1/90 × 15.05 inch = 0.167 inch.

As gears E and F are on the same shaft, one faulty tooth of the former also produces a periodicity having 2.5 inch wavelength. However, the width of the fault due to one faulty tooth on gear E will be =1/89 × 15/90 × 15.05 inch = 0.028 inch. Similarly, one faulty tooth of D or change wheel

TABLE 10.1   Wavelength of Fault due to Single Faulty Tooth

Faulty Gear	Wavelength of Fault (inch)
G	15.05 (corresponds to one revolution of gear G)
F	2.5 (corresponds to one revolution of gear F)
E	2.5 (corresponds to one revolution of gear E)
D	0.68 (corresponds to one revolution of gear D)
Change wheel	0.68 (corresponds to one revolution of change wheel)
A or B	24.35/CW

TABLE 10.2   Width of Fault due to Single Faulty Tooth

Faulty Gear	Calculation	Width of Fault (inch)
F or G	15.05/90	0.167
D or E	$(15/89) \times (15.05/90)$	0.028
B or C	$(24/CW) \times 0.028$	0.676/CW
A	$(36/24) \times (0.672/CW)$	1.01/CW

produces periodicity having wavelength = $24/89 \times 15/90 \times 15.05 = 0.676$ inch. One faulty tooth of A or B produces periodicity having wavelength = $36/CW \times 24/89 \times 15/90 \times 15.05 = 24.35/CW$ inch. Tables 10.1 and 10.2 summarise the wavelength and width of the faults due to one faulty tooth.

Figure 10.6 depicts the wavelength and width of the faults generated due to one faulty tooth on wheel G and F.

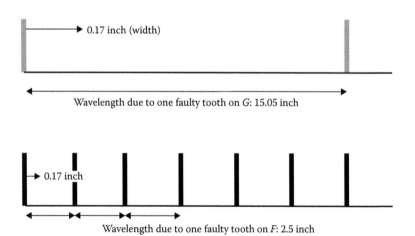

FIGURE 10.6   Wavelength and width of the faults due to one faulty tooth on gears G and F.

10.1.4.2 Case II: All Teeth of Any One Gear Are Faulty

Either G or F

$$\lambda = \frac{15.05}{90} = 0.167 \text{ inch}$$

Either E or D

$$\lambda = \frac{1}{89} \times \frac{15}{90} \times 15.05 = 0.028 \text{ inch}$$

Either CW or B

$$\lambda = \frac{1}{CW} \times \frac{24}{89} \times \frac{15}{90} \times 15.05 = \frac{0.676}{CW} \text{ inch}$$

Gear A

$$\lambda = \frac{1}{24} \times \frac{36}{CW} \times \frac{24}{89} \times \frac{15}{90} \times 15.05 = \frac{1.015}{CW} \text{ inch}$$

10.1.4.3 Case III: Any One Gear Is Eccentric

Either G or take-up roller

$$\lambda = 15.05 \text{ inch}$$

Either E or F

$$\lambda = \frac{15}{90} \times 15.05 = 2.5 \text{ inch}$$

Either CW or D

$$\lambda = \frac{24}{89} \times \frac{15}{90} \times 15.05 = 0.676 \text{ inch}$$

Either A or B

$$\lambda = \frac{36}{CW} \times \frac{24}{89} \times \frac{15}{90} \times 15.05 = \frac{24.35}{CW} \text{ inch}$$

It is noted from these calculations that in many cases defective gears are liable to produce dangerous periodicities for seven-wheel take-up.

10.1.5 Shirley Take-Up

Shirley take-up is a continuous take-up motion as depicted in Figure 10.7. Gear A is driven continuously by chain and sprocket at one quarter of the loom speed, that is picks per minute. Gear A drives change wheel (*CW*) through the career gear B. A single worm (D) on the same shaft of *CW* drives a worm gear E which is mounted on the take-up roller shaft. The use of worm and worm gear combination causes drastic reduction in rotational speed in one step which makes Shirley take-up system different from five- and seven-wheel take-up systems. Besides, the rotation of take-up roller is also continuous. The circumference of take-up roller is 10 inch.

The amount of fabric taken up for each pick corresponds to the pick spacing and can be calculated as

$$\text{Pick spacing} = \frac{1}{4} \times \frac{60}{CW} \times \frac{1}{150} \times 10 = \frac{1}{CW} \text{ inch}$$

Therefore

$$\text{Picks per inch or } PPI = CW$$

Thus the number of teeth on the change wheel is equal to the picks per inch. Calculation of wavelength of periodicities for different cases for Shirley take-up is shown in the following text:

Case I: All teeth of any one gear are faulty:
Either E or D

$$\lambda = \frac{1}{150} \times 10 = 0.067 \text{ inch}$$

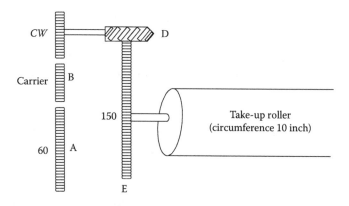

FIGURE 10.7   Shirley take-up motion.

TABLE 10.3   Wavelength of Faults in Shirley Take-Up Motion

Faulty Gear	Calculation	Wavelength of Fault (Inch)
Periods due to eccentricity		
Take-up roller or worm gear	—	10
Worm *D* or change wheel	$(1/150) \times 10$ inch	0.067
Gear *A*	$(60/CW) \times 0.067$ inch	$4/CW$
Periods due to faulty gear		
Worm *D* or worm wheel *E*	$(1/150) \times 10$ inch	0.067
A or *B* or change wheel	$(1/CW) \times 0.067$ inch	$0.067/CW$

Either A or B or change wheel

$$\lambda = \frac{1}{CW} \times \frac{1}{150} \times 10 = \frac{0.067}{CW} \text{ inch}$$

Case II: Any one wheel is eccentric:
Either E or take-up roller

$$\lambda = 10 \text{ inch}$$

Either D or change wheel

$$\lambda = \frac{1}{150} \times 10 = 0.067 \text{ inch}$$

Gear A

$$\lambda = \frac{60}{CW} \times \frac{1}{150} \times 10 = \frac{4}{CW} \text{ inch}$$

Table 10.3 summarises the wavelengths of various faults.

From this calculation it is clear that there are no periodicities of wavelength between 1/8 and 10 inch, so the risk of dangerous periodicities is eliminated in Shirley take-up motion (Marks and Robinson, 1976).

10.2 LET-OFF MOTION

10.2.1 Objectives

The objectives of let-off motion is to maintain the free length of warp within specified limits and to control the warp tension by means of supplying the warp at a correct rate to the weaving zone.

10.2.2 Classification of Let-Off

Let-off motion is classified under three categories as given here.

1. Negative

2. Semi-positive

3. Positive

In the case of negative let-off, warp is pulled from the weaver's beam against a slipping-friction system. In semi-positive let-off, the weaver's beam is rotated positively by a beam-driving mechanism. However, the extent of weaver's beam rotation is determined by the warp tension. A greater increase in warp tension causes greater rotation of the weaver's beam and vice versa. In the case of semi-positive let-off, the movement of the warp tension–sensing device (generally backrest) is utilised to rotate the weaver's beam. In this case, the tension-sensing device and beam-driving mechanism are not completely independent. For positive let-off system, weaver's beam is rotated by driving mechanism at a controlled rate in order to maintain a constant length of warp between the weaver's beam and cloth fell. In this case, the means of applying warp tension is separated from the beam-driving mechanism.

10.2.2.1 Negative Let-Off

The negative let-off mechanism is illustrated in Figure 10.8. In this case, the warp is pulled off the weaver's beam and warp tension is governed by the friction between chain and the beam ruffle. The chain makes some wrap over the ruffle. Slack side of the chain is attached with the machine frame, whereas the tight side is attached with the weight lever. The lever has fulcrum at one end (H) with the machine frame. The other end carries dead weight (W). Two such systems are attached at the two sides of the loom.

Let R is the radius of warp layers on the weaver's beam

r is the beam ruffle radius

T_t is the tension in the chain on tight side (attached with the weight lever)

T_s is the tension in the chain on slack side (attached with the machine frame)

W is dead weight

x is the distance between fulcrum point and chain on tight side (fixed)

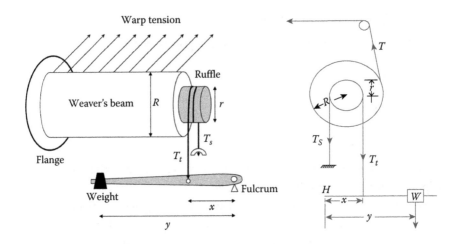

FIGURE 10.8 Negative let-off motion.

y is the distance between fulcrum point and weight (variable)

T is the tension in the warp sheet (variable)

Taking moments about the beam centre

$$T \times R = \left(T_t - T_s\right) r \tag{10.1}$$

Now

$$\frac{T_t}{T_s} = e^{-\mu\theta} \tag{10.2}$$

where
  μ is the coefficient of friction between the chain and the beam ruffle
  θ *is* angle of wrap in radian made by the chain on beam ruffle

So

$$TR = T_t \left(1 - \frac{T_s}{T_t}\right) r = T_t \left(1 - e^{-\mu\theta}\right) r \tag{10.3}$$

Now, taking moments about the fulcrum H of the lever,

$$T_t x = W y$$

or

$$T_t = Wy/x$$

So

$$TR = \frac{Wy}{x}\left(1-e^{-\mu\theta}\right)r \quad \text{or} \quad T = W\times\frac{r}{R}\times\frac{y}{x}\left(1-e^{-\mu\theta}\right) \qquad (10.4)$$

Therefore

$$T \propto \frac{y}{R} \qquad (10.5)$$

Equation 10.5 shows that to achieve a constant warp tension, the ratio y/R should be kept constant. Thus as beam radius R reduces, the distance y must be reduced by moving the weight towards the fulcrum H at regular interval to balance the warp tension. For example, if the beam radius decreases by 25%, the distance y must be reduced by 25% to maintain a constant warp tension. As shown in Figure 10.9, the warp tension is maintained within a small range from full beam to empty beam by shifting the weights at some intervals. The maximum permissible warp tension is indicated by the straight line (T_{max}). As soon as the warp tension reaches T_{max},

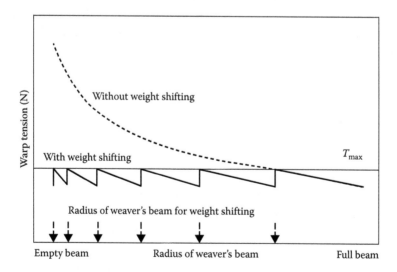

FIGURE 10.9   Warp tension versus beam radius.

TABLE 10.4   Diameter of Warp Beam for Weight Shifting

	Full Beam Diameter (cm)			
	100		60	
Permissible increase in tension (%)	25	33	25	33
Diameter for 1st change (cm)	80	75.2	48	45.1
Diameter for 2nd change (cm)	64	56.6	38.4	33.9
Diameter for 3rd change (cm)	51.2	42.5	30.7	25.5

the dead weight is shifted to reduce the warp tension instantaneously. It is also noted that the frequency of weight shifting increases as the beam gets exhausted. This is ascribed to the asymptotic relationship between warp tension and radius of warp on the beam. If the weight is not shifted, then the warp tension will increase continuously with the reduction of beam diameter. This is indicated by the broken line.

Let us consider one hypothetical example. The initial diameter of a weaver's beam is 100 cm. The allowable increase in warp tension is 25% of nominal level. So, when the beam diameter will be 80 cm, the warp tension will increase by 25%. The weaver will adjust the position of the weight so that the tension will come back to nominal level. In the second step, when the beam diameter will reduce to 64 cm, the warp tension will again increase by 25%. Therefore, first weight shifting will be done after a 20 cm reduction in beam diameter, whereas the second weight shifting will happen after a 16 cm reduction in beam diameter. Besides, the length of warp sheet delivered by the weaver's beam in its one revolution is also more when the beam diameter is more. So, as the beam weaves down, the shifting of weight will be more frequent. Table 10.4 presents the diameters of warp beam at which weight shifting has to be done. It is observed that if permissible increase in tension is more, weight shifting is delayed as expected.

10.2.2.2 Semi-Positive Let-Off

The schematic view of a semi-positive let-off system is shown in Figure 10.10. The warp sheet passes over the backrest which is movable or floating type. The backrest is carried by two long levers (swing levers) at the two sides of the loom. If the warp tension increases, the backrest is depressed or lowered. So the tail end of the swing lever is raised. This upward movement of the lever is transmitted to the pawl and the latter rotates the ratchet. As a result the weaver's beam is rotated and warp sheet is released to lower down the tension. On the other hand, if the warp tension is reduced, the backrest is raised and the tail end of the swing lever

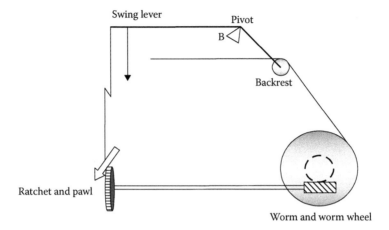

FIGURE 10.10   Semi-positive let-off.

is lowered. However, the pawl fails to rotate the ratchet in this case as the former is disengaged by a sheath which locks the ratchet.

The system for applying warp tension in semi-positive let-off is shown in Figure 10.11. The weight lever pivoted at A is carrying the dead weight W. The swing lever, which has two limbs of length c and d, is pivoted at B. The weight lever is experiencing a clockwise torque due to the weight W. Therefore, the swing lever experiences an anti-clockwise torque. So the backrest applies force F on the warp sheet.

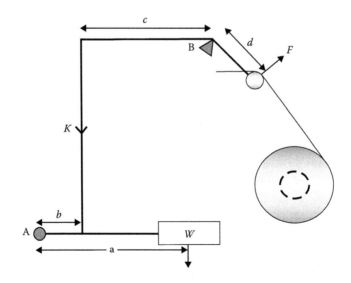

FIGURE 10.11   Warp tensioning in semi-positive let-off.

At equilibrium:

$$F \times d = K \times c$$

So

$$K = \frac{F \times d}{c}$$

Moreover

$$K \times b = W \times a$$

So

$$\frac{F \times d}{c} \times b = W \times a$$

or

$$F = W \times \frac{a \times c}{b \times d} \tag{10.5}$$

Considering an additional fixed force P due to the weight of the levers, this expression becomes

$$F = W \times \frac{a \times c}{b \times d} + P. \tag{10.6}$$

The force F, which the backrest applies on the warp sheet, will vary depending on the effective diameter of the warp layer on the weaver's beam. This will happen as the angle between the two limbs of the warp sheet around the backrest changes from the start to the finish of the weaver's beam.

10.2.2.3 Positive Let-Off

In the case of positive let-off, the warp tension is controlled by a mechanism which drives the warp beam at a correct rate. In most of the positive let-off systems, the backrest is not fixed but floats. It acts as a warp tension–sensing mechanism. As the tension in the warp increases, the backrest is depressed. A Hunt positive let-off motion is illustrated in Figure 10.12. There are two

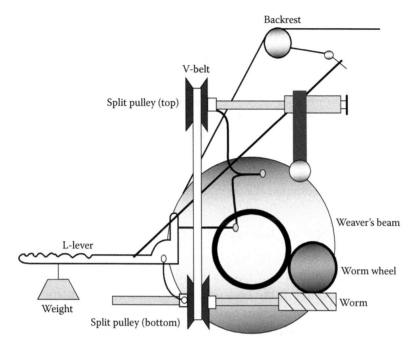

FIGURE 10.12   Hunt positive let-off motion.

split pulleys made out of *V*-pulley. Motion from the crankshaft moves the top split pulley via a worm and worm wheel. The top pulley in turn drives the bottom pulley through a belt. As the tension on the warp increases, the backrest goes down and the *L*-type lever with weight lowers the diameter of the bottom pulley and essentially increases the diameter of the top pulley through the necessary linkages. Now the bottom pulley moves at a faster rate than it was earlier. As a result, the connecting worm to the beam drive moves faster to deliver extra warp to reduce the warp tension.

10.3 AUXILIARY OR STOP-MOTIONS

These motions are used to stop the loom in the following cases.

- Shuttle trapping (warp protecting motion)
- Weft break (weft stop-motion)
- Warp break (warp stop-motion)

The classification of stop-motions is shown in Figure 10.13.

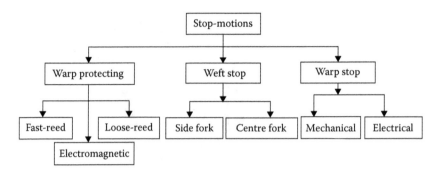

FIGURE 10.13   Classification of stop-motions.

10.3.1 Warp Protecting Motion

Warp protecting motion protects the warp sheet when the shuttle gets trapped inside the shed or it ricochets and comes back into the shed due to improper checking. If the beat-up is performed in this situation, then a large number of warp yarns will break and even the shuttle may get damaged. The role of warp protecting motion is to stop the loom, before beat-up, in such cases.

10.3.1.1 Fast-Reed Warp Protecting Motion

The working principle of fast-reed motion is shown in Figure 10.14 (Lord and Mohamed, 1982). The swell used for shuttle checking is attached with the back wall of the shuttle box. When the shuttle reaches the shuttle box safely, the swell retards the shuttle and in the process the swell is displaced towards the right-hand side. Therefore, the finger–dagger assembly rotates clockwise. Thus when the dagger moves forward with the sley, it clears the frog which is fixed on the loom frame. If the shuttle is trapped inside the shed, then the dagger hits the frog when the sley assembly moves towards the left-hand side (front centre) for performing the beat-up. The frog is connected to the starting handle of the loom. The loom is stopped immediately with a loud sound that is known as 'bang-off'.

10.3.1.2 Loose-Reed Warp Protecting Motion

The working principle of loose-reed motion is shown in Figure 10.15. The reed is supported by two baulks. The top baulk is fixed, whereas the bottom one is loose but remains in position due to spring pressure. If the

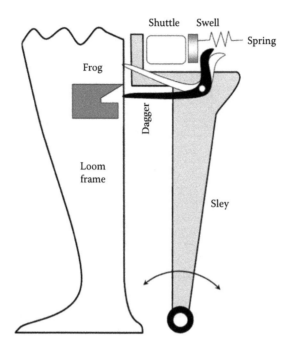

FIGURE 10.14 Fast-reed warp protector motion.

shuttle is trapped inside the shed, the reed experiences pressure when it moves towards the cloth fell for performing the beat-up. This pressure rotates the bottom baulk support and the entire assembly in the anticlockwise direction as shown on the left-hand side of Figure 10.15. So the dagger moves up to the level of the hitter which is fixed on the loom. The loom is stopped as the dagger hits the hitter. When the shuttle reaches the destination properly, a situation similar to what is depicted on the right-hand side of Figure 10.15 is created. The dagger passes beneath the hitter and the bottom baulk of the loose-reed is supported by the finger for effective beat-up.

The comparison between fast-reed and loose-reed warp protector motions is presented in Table 10.5.

10.3.1.3 Electromagnetic Warp Protecting Motion

In this case, the shuttle and the crankshaft wheel carry magnets to create electromagnetic induction in coils. If the shuttle does not arrive in the shuttle box, then the sequence of electrical signal generation is violated and the loom is stopped.

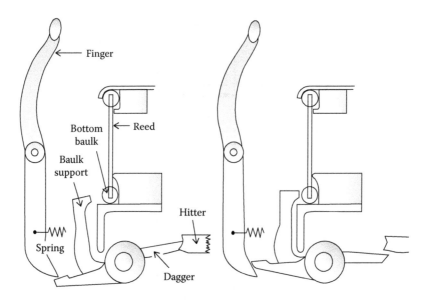

FIGURE 10.15   Loose-reed warp protector motion. (From Marks, R. and Robinson, A.T., *Principles of Weaving*, The Textile Institute, Manchester, UK, 1976.)

TABLE 10.5   Comparison between Fast-Reed and Loose-Reed Motions

Fast-Reed	Loose-Reed
Shuttle should contact the swell at 250° and should displace the swell completely by 270°.	There is such limitation of timing.
Less time is available for shuttle flight, so limitation is imposed on higher loom speed.	More time is available for shuttle flight. Higher loom speed can be attained.
For same loom speed, shuttle velocity will be relatively higher.	For same loom speed, shuttle velocity will be relatively lower.
Strain in warp yarns is lower in the case of shuttle trapping.	Strain in warp yarns is higher in the case of shuttle trapping.

10.3.2 Warp Stop-Motion

Warp stop-motion stops the loom in the event of an end break. The system is activated by the lightweight metallic drop wires which have profiled shape. Two such drop wires or droppers are shown in Figure 10.16. The large slot at the top is for the movement of the reciprocating bars which are used in both mechanical and electrical warp stop-motions. Drop wires having design (a) can be used when a single end is passed through a drop wire during the drawing in operation. Drop wires having design (b) can be used after the beam gaiting as it has an open-ended hole.

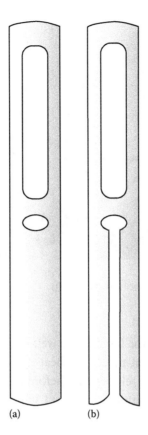

FIGURE 10.16   Drop wires (a) Closed hole type (b) Open hole type.

In the case of mechanical warp stop-motion, one reciprocating bar moves between two stationary bars. The bars have profiles like step waves. The sideways movement of the centre bar is equal to the width of a step. In the case of an end break, the drop wire will lose support from the yarn and will fall due to gravity. If it falls to the lowest possible height, then the reciprocating movement of the centre bar will be thwarted and the loom is stopped. In the case of electrical stop-motion, the drop wire acts as an element that makes or breaks an electrical circuit. In the case of an end break, the drop wire will complete an electrical circuit and activate a solenoid. The solenoid will attract a bar which will be hit by the knock-off lever. As a result the bar will disengage the starting handle through some other levers. Figure 10.17 shows the warp stop-motion mounted on a loom.

FIGURE 10.17   Warp stop-motion on a loom.

10.3.3 Weft Stop-Motion

Weft stop-motions stop the loom if the weft carried by the shuttle is broken. It is a very important motion as beat-up without a pick will necessitate adjustment of cloth fell position before the restart of the loom. The problem of cloth fell position adjustment will be relatively lower in the following cases.

- Coarser yarn count
- Higher yarn hairiness
- Higher pick density

10.3.3.1 Side Weft Fork Motion

Side weft fork motion operates on the left-hand side of the loom at the vicinity of starting handle. When the shuttle reaches the shuttle box inserting an unbroken pick, the trail of pick pushes the lower end of the fork as the sley moves forward as depicted in Figure 10.18 (Lord and Mohamed, 1982). This creates an anti-clockwise movement in the fork according to Figure 10.18. However, the movement of the fork will be clockwise according to Figure 10.19. From Figure 10.19, it can be understood that the notched hammer moves towards the front of the loom once in two picks as it gets motion from a cam mounted on the bottom shaft. Under normal circumstances, the movement of the fork created by the push exerted by

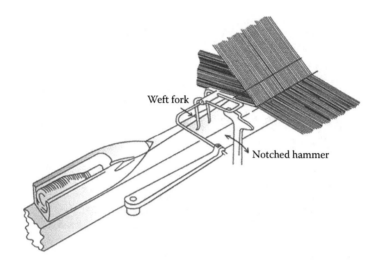

FIGURE 10.18 Side fork and notched hammer. Reprinted from *Weaving: Conversion of Yarn to Fabric*, 2nd edn., Lord, P.R. and Mohamed, M.H, Copyright 1982, with permission from Elsevier.

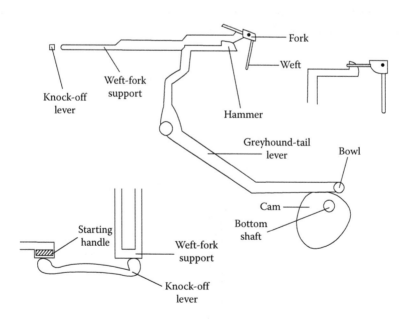

FIGURE 10.19 Side view of side weft fork motion. (From Marks, R. and Robinson, A.T., *Principles of Weaving*, The Textile Institute, Manchester, UK, 1976.)

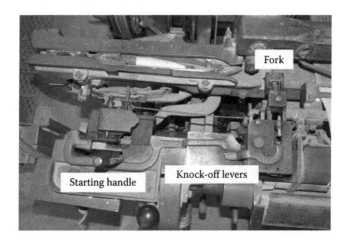

FIGURE 10.20   Top view of side weft fork motion.

the pick clears the upper end of the fork from the notched hammer when the latter moves towards the front of the loom. Thus the loom continues to run. In the case of a weft break, the upper end of the fork is caught by the notch of the hammer. So, when the hammer moves towards the front of the loom, the weft fork support pushes the knock-off lever and the latter dislocates the starting handle to stop the loom. Figure 10.20 presents the top view of the side weft fork system mounted on a loom. Side weft fork

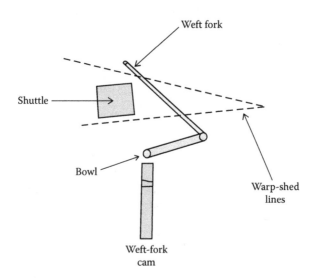

FIGURE 10.21   Centre weft fork motion.

system can detect weft break after the insertion of one or two missing picks. This happens as the side weft fork system is placed on the left-hand side of the loom and it can stop the loom if the pick is missing when shuttle has been propelled from the right-hand side of the loom.

10.3.3.2 Centre Weft Fork Motion

This problem of side weft fork system can be mitigated by using centre weft form motion which is mounted near the middle of the loom. It checks for the weft break at every pick and stops the loom before the beat-up in case there is any. Thus the centre weft fork motion is more efficient than the side weft fork motion. Figure 10.21 shows the centre weft fork, which is housed in a slot on the sley. The fork rotates clockwise to make a clear passage for the shuttle. This is done by the left sideways movement of the weft fork cam. In the presence of a pick, the fork is supported by the former when the sley moves forward for the beat-up. In the case of a weft break, the fork loses the support and thus the weft fork bowl will be lowered and trapped in a notch restricting the movement of a rod which finally creates the loom stoppage.

NUMERICAL PROBLEMS

10.1 A simple weight lever–based negative let-off system is attached at each end of a weaver's beam. It is provided with weights of 500 N at each side. The leverage of the system (y/x ratio) is 5:1. The radius of full beam is 80 cm and the ruffle radius is 20 cm. If the ropes are given 1.5 warps around the ruffles and the coefficient of friction between rope and ruffle is 0.20, then determine the warp tension at the slipping point.

Solution:

$$\text{Warp tension} = T = \frac{r}{R} \times \frac{y}{x} \times W(1 - e^{-\mu\theta})$$

Here

Ruffle radius $= r = 20$ cm and beam radius $= R = 80$ cm.

$$\theta = 2\pi \times 1.5 = 3\pi$$

$$\mu = 0.2$$

$$y/x = 5:1$$

$$W = 500 \text{ N}$$

So

$$T = \frac{20}{80} \times \frac{5}{1}(1 - e^{-0.2 \times 3\pi}) \times 500$$

$$= 1.25 \times (1 - 0.152) \times 500$$

$$= 530 \text{ N}$$

Considering two sides of the loom

The warp tension will be $= 530 \times 2 = 1060$ N

So the warp tension at the slipping point is 1060 N.

10.2 A 75 Stockport reed (number of dents/2 inch) of 2 m width is being used on a loom. The reed plan is two ends in a dent. If the weft crimp is 6% then calculate the ends per cm in the fabric.

Solution:
There are 75 reed dents in 2 inch. Two ends are passing through one dent.
Therefore

$$\text{Total ends in the reed} = \frac{200}{2.54} \times \frac{75}{2} \times 2$$

$$= 5905.$$

$$\text{Crimp} = \frac{L_y - L_f}{L_f} = \frac{6}{100} = 0.06$$

where L_y and L_f are width of warp sheet and fabric, respectively.
So

$$\frac{L_y}{L_f} = 1.06$$

or

$$L_f = \frac{L_y}{1.06} = \frac{\text{Reed width}}{1.06} = \frac{200}{1.06} = 188.7 \text{ cm}$$

So 5905 ends will be accommodated in 188.7 cm width of the fabric. Therefore

$$\text{ends/cm} = \frac{5905}{188.7} = 31.3$$

So ends per cm is 31.3.

10.3 A shuttle loom, equipped with a seven-wheel take-up motion, is running at 180 picks per minute. The change wheel is having 100 teeth. If the reed width is 200 cm and width-wise fabric contraction is 8% after the temple, calculate the production capacity of the loom in m²/hour.

Solution:
In the case of seven-wheel take-up motion

$$\text{Picks/inch} = \frac{\text{Number of teeth on change wheel}}{1.015}$$

So

$$\text{Picks/cm} = \frac{\text{Number of teeth on change wheel}}{1.015 \times 2.54}$$

$$= \frac{100}{1.015 \times 2.54} = 38.79$$

So

Delivery speed of the fabric (m/hour)

$$= \frac{\text{Number of picks inserted/hour}}{\text{Picks/cm}} \times \frac{1}{100}$$

$$= \frac{180 \times 60}{38.79 \times 100} = 2.784$$

$$\text{Contraction} = \frac{L_y - L_f}{L_y} = 0.08$$

where L_y and L_f are width of warp sheet and fabric, respectively. The warp sheet width is 200 cm. However, the width of the fabric will be reduced after the temple due to the crimp generation in weft.

So

$$\frac{200 - L_f}{200} = 0.08$$

or

$$L_f = 184 \text{ cm} = 1.84 \text{ m}$$

So

The fabric production in m²/hour

$$= \text{Delivery speed of the fabric (m/hour)} \times \text{Fabric width}$$

$$= 2.784 \times 1.84 = 5.12.$$

So production capacity of the loom is 5.12 m²/h.

10.4 Determine the relationship between picks per inch (*PPI*) and the number of tooth on change wheel (*CW*) for the (Picanol) take-up system shown in Figure 10.22. Calculate the wavelength of periodic faults which will result from the mechanical defects of the system.

Solution:

The gear *A* moves by one tooth per pick.
So

$$\text{Pick spacing} = \frac{1}{41} \times \frac{21}{42} \times \frac{42}{CW} \times \frac{16}{63} \times \frac{15}{70} \times 35.85 \text{ cm} = \frac{1}{CW} \text{ cm.}$$

So

$$\text{Picks/cm} = CW.$$

The calculations for wavelengths are shown in Tables 10.6 and 10.7.

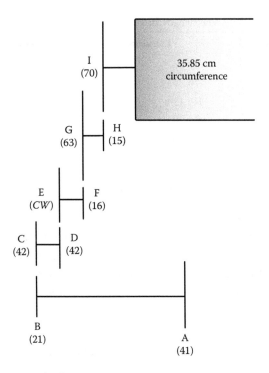

FIGURE 10.22   Picanol take-up system.

TABLE 10.6   Wavelength for Eccentric Elements

Eccentric Element	Calculation	Wavelength (cm)
Take-up roller or I	—	35.85
G or H	$\dfrac{15}{70} \times 35.85$	7.68
E (change wheel) or F	$\dfrac{16}{63} \times \dfrac{15}{70} \times 35.85$	1.95
C or D	$\dfrac{42}{CW} \times \dfrac{16}{63} \times \dfrac{15}{70} \times 35.85$	$\dfrac{81.94}{CW}$
A or B	$\dfrac{21}{42} \times \dfrac{42}{CW} \times \dfrac{16}{63} \times \dfrac{15}{70} \times 35.85$	$\dfrac{40.97}{CW}$

TABLE 10.7   Wavelength for Faulty Gears (All Teeth Worn Out)

Faulty Gear	Calculation	Wavelength (cm)
I or H	$\dfrac{35.85}{70}$	0.512
F or G	$\dfrac{1}{63} \times \dfrac{15}{70} \times 35.85$	0.122
D or E	$\dfrac{1}{CW} \times \dfrac{16}{63} \times \dfrac{15}{70} \times 35.85$	$\dfrac{1.95}{CW}$
B or C	$\dfrac{1}{42} \times \dfrac{42}{CW} \times \dfrac{16}{63} \times \dfrac{15}{70} \times 35.85$	$\dfrac{1.95}{CW}$

10.5 A loom is running with negative let-off motion. The full and empty diameter of weaver's beam is 60 and 20 cm, respectively. The weaver does not want the warp tension variation to exceed by 20% during the weaving. How many times the weight has to be shifted during the weaving?

Solution:

$$\text{Warp tension} = T = \frac{r}{R} \times \frac{y}{x} \times W(1 - e^{-\mu\theta})$$

Here full beam diameter = R_1 = 60 cm and empty beam diameter R_f = 20 cm. Let the final tension in the warp be T_f and initial tension in the warp be T_1. Tension in the warp varies inversely with the weaver's beam diameter.

If the weight is not shifted during weaving, then

$$\frac{T_f}{T_1} = \frac{R_1}{R_f} = \frac{60}{20} = 3$$

So warp tension will increase by 300%.

Permissible increase in tension is only 20%. This implies that the maximum warp tension can be 1.2 times of initial tension. So

$$T_f = 3T_1 = 1.2^n \times T_1$$

So

$$1.2^n = 3$$

or

$$n = 6 \text{ times.}$$

As warp tension is inversely proportionate with beam diameter, the latter should reduce by a factor of $1/1.2 = 0.833$ to increase the warp tension by 20%. Diameters at which the weight has to be shifted to bring down the tension to T_1 are as follows.

1st change at $60 \times 0.833 = 50$ cm

2nd change at $50 \times 0.833 = 41.67$ cm

3rd change at $41.67 \times 0.833 = 34.72$ cm

4th change at $34.72 \times 0.833 = 28.93$ cm

5th change at $28.93 \times 0.833 = 24.1$ cm

6th change at $24.1 \times 0.833 = 20$ cm

So the weight has to be shifted six times. The last one may be avoided as the beam has become empty (20 cm).

REFERENCES

Lord, P. R. and Mohamed, M. H. 1982. *Weaving: Conversion of Yarn to Fabric*, 2nd edn. Merrow, UK: Merrow Technical Library.

Marks, R. and Robinson, A. T. C. 1976. *Principles of Weaving*. Manchester, UK: The Textile Institute.

Index

9 780367 574192